CW01085298

SCANIA at work:

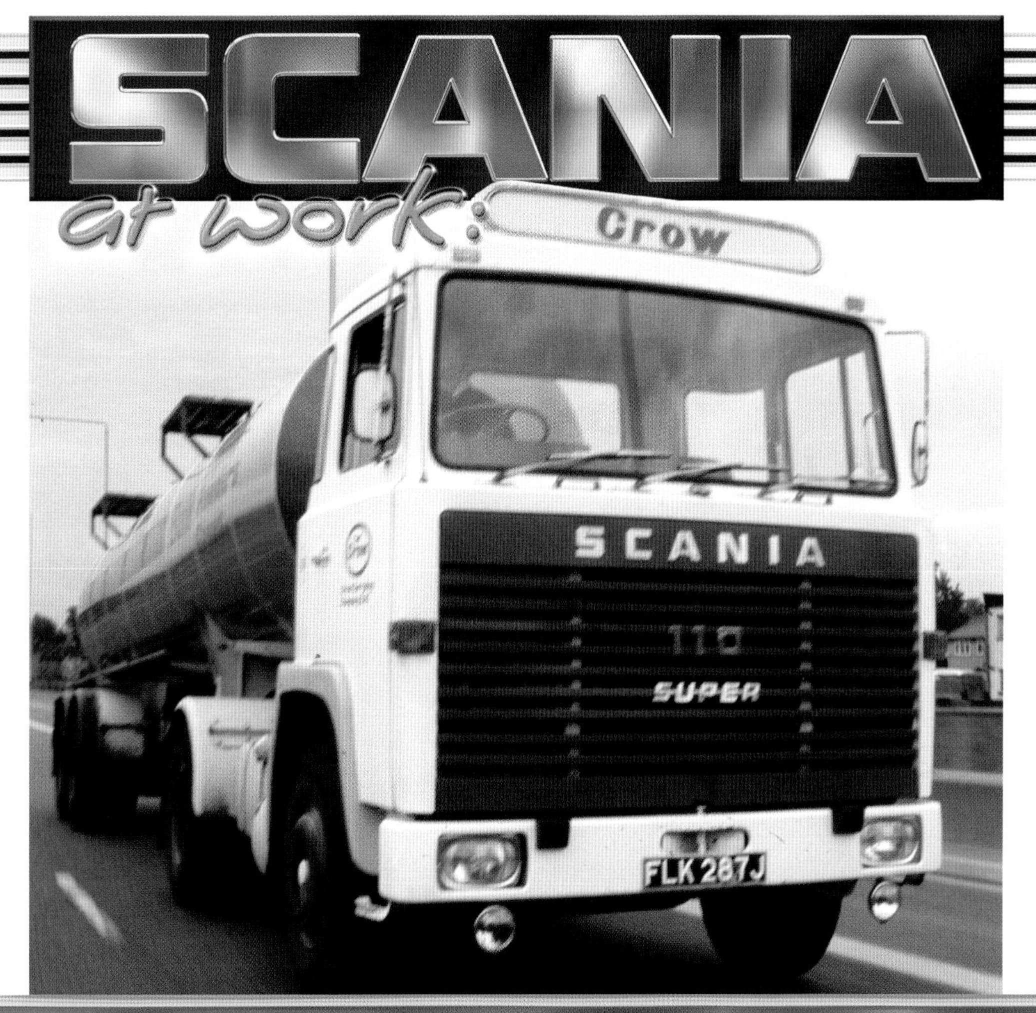

LB110, 111, 140 & 141

Patrick W Dyer

ACKNOWLEDGEMENTS

As ever, there are many people who helped in bringing this project to fruition. However, particular thanks must go to the following: Jackie Burr, Phil Sampson, Sarah Tennent, Chris Love, Adrian Cypher, Marcus Lester, David Wakefield, Del Roll, Brian Weatherley & *Commercial Motor,* John W Henderson, Leif Ohrn, Steve Cole, Dick Phillips, Ron Smith, Marie-Louise Lindholm, Hans-Ake Danielsson, George Bennett, Colin Dunn (West Pennine Trucks), Sue Chapman and, of course, my long-suffering wife, Linda.

ABOUT THE AUTHOR

Patrick Dyer, born in 1968, grew up during one of the most notable and exciting periods of development for heavy trucks and also the last of the real glory days for trucking as an industry. This is reflected in his subject matter. His first book covered the revolutionary Volvo F88 and F89; this new title deals with the equivalent products from Scania. Although Patrick's day job is in motor sport, he holds a Class One licence and drives whenever the opportunity arises. He is also the proud owner of a 1983 Volvo F12, which he is returning to its original condition with the help of long-term friend, and noted classic vehicle restorer, Ashley Pierce.

DEDICATION

This one's for Linda and Jess. Well done for putting up with my 'Truck' habit!

First published 2009

ISBN 978-1-905523-99-3

A catalogue record for this book is available from the British Library

Published by
Old Pond Publishing Ltd
Dencora Business Centre,
36 White House Road
Ipswich IP1 5LT
United Kingdom

www.oldpond.com

Front cover photograph
'As the Crow flies': among the first foreign vehicles purchased by the Crow Carrying Company were a couple of LB110s fitted with the turbocharged DS11 engine. With 260 bhp, power steering, a synchromesh gearbox, sprung seats and a comfortable cab they proved very popular with the drivers. FLK 287J, being a 1971 example, benefited from the more normal mirror arm mountings and the new, larger SCANIA lettering and badges. *(Scania AB)*

Back cover photograph
The LB141 represented the pinnacle of the Scania LB range. With an improved version of the DS14 engine that produced 375 bhp Scania regained the crown of the most powerful truck in Europe. The LB range was finally replaced by the Series Two Scanias in 1981, which, although featuring new cabs and chassis, used further developed examples of the same basic engines, gearboxes and axles as the old models. *(Scania AB)*

Cover design and book layout by Liz Whatling
Printed and bound in China

Contents

Foreword

By George Bennett

Editor, Truck *magazine 1987-89 & 1990-97*

In his first book, Patrick Dyer described the history of Volvo's iconic F88, but the subject of this book, Scania's LB range – better known as the 110/140 and 111/141 – is truly legendary. In the 1970s nearly every driver acknowledged the Scania 141 as the King of the Road, even those committed, as I was with Volvo and Daf, to other marques. The 141 was the most powerful truck of its day, but it also figured in some of the more glamorous – and in some cases more swashbuckling – operations such as long-haul fridge transport from Ireland and Scotland. The 141 set a standard for other manufacturers to match and it created a premium image for Scania that still persists some thirty years later.

This was no fluke, as Patrick describes in admirable detail. The evolution of the L75 into the LB76 and on to the LB110 and LB111 (and their V8 stable-mates) was one of careful attention to trends in the transport market and equally thorough commitment to engineering detail and driver comfort.

Scania and Volvo were particularly influential in the British truck and transport industries because they were the first continental manufacturers to sell trucks in Britain in significant numbers, thanks to the fact that both Britain and Sweden were in the European Free Trade Area (EFTA) long before both countries joined the EEC. The two Swedish marques thus enjoyed favourable import tariffs that were unavailable to other European manufacturers. They were also the advance guard of a continental invasion that signalled the decline of the previously invincible British truck industry, and its eventual demise.

The Scania 110 was such an impressive beast that I can still recall the first time I saw one, even before I passed my HGV test, when a solo tractor unit passed across a junction in front of me as I was walking home one evening. The sheer height of the day-cab was startling and it took me a while to realise what it was, though I saw immediately that it was revolutionary compared with the Leylands, Atkis and ERFs I was used to seeing.

My first Scania drive was in a 111 operated by Cadwallader of Oswestry. Caddy's were mainly a Volvo firm, but they bought a few Scania 111s for their French subsidiary, and later a few right-hand-drive examples. Although my regular truck was a Volvo F88, followed by a Daf 2800 and a Volvo F10, for reasons I can no longer remember I found myself at the wheel of a 111 en route from Oswestry to Poole and the ferry to Cherbourg. The Scania cab was broad and spacious compared with the F88, and rivalled the 2800 for interior space. It had a reassuring solidity, too, compared to the more skinny panels of the big Daf.

Although I loved the commanding view from the Scania, the truck had its quirks. Chief among these was the ten-speed range-change gearbox (we drivers didn't know it was called the GR860). Its gearshift was designed for left-hand drive operation but in a right-hooker the shift through the range-change from fifth to sixth involved drawing the stick back as you pushed it right across the gate away from you to find sixth. It was so awkward that for most of the time I ignored sixth altogether and drove it like a nine-speed with a gaping step between fifth and seventh. Comparing notes later with regular 111 drivers, I found this was common practice, though what the price was in terms of clutch wear I never discovered.

More than twenty years later I found myself driving a 111 in a twenty-fifth anniversary group test for *Truck* magazine, in October 1999. There to welcome me back was the awkward gearshift, and once again I spent most of the time ignoring sixth gear altogether. Of course, no one warned me about the gearshift back in the 1970s. Following standard practice among hauliers in those days, it was simply assumed that I knew how to drive the 111 and of course I did, up to a point. But I wasn't told about the Scania philosophy of letting the engine lug. It was poignant to read in this book how carefully Scania pioneered and promoted the 'drive on green' philosophy and to ponder how few operators passed on this information to drivers, let alone trained them to use it. If I had any idea of the significance of the green band in the rev counter it was from reading Pat Kennett in *Truck* magazine, and not from in-company training.

Most drivers agreed that Volvo and Scania were the top marques in those days but in terms of interior space and in-cab noise levels the Scania 111/141 had the better of the Volvo F88. It was also cooler in summer because the engine cover was less intrusive. And if you needed to get to the engine there was no contest. The F88 cab had to be tilted by hand, which was no joke if it was laden with long-haul driver's gear, as I discovered several times to my cost. On any of these occasions I would have cheerfully swapped to a Scania there and then, if only to enjoy the hydraulic cab-tilt pump.

Nevertheless, as so often happens, Volvo leapfrogged Scania with the launch of the F10 which, although it wasn't as roomy as we Volvo drivers hoped, did offer the driver a better ride.

I discovered this by chance on a run to Bucharest in 1981, running with a Scania 111 driven by a Cadwallader colleague known for his speedy driving. Speedy or not, he kept falling behind, and finally I stopped to ask him why. It was all right for me, he said, in my well-sprung F10 cab, but he was being shaken to pieces in the older and rubber-sprung Scania, and he couldn't maintain the pace. We slowed down, but not a lot.

Whatever the relative technical merits of the mainstream models from the two Swedish marques, no other contemporary truck could match the reputation of the Scania 141 with its 375 bhp V8. As Patrick points out, following its launch, 'the model went on to build a legendary reputation for performance and reliability with indestructible engines that were capable of staggering mileages even under the harshest of conditions.' And even under relatively benign western European conditions, the 141 sometimes operated under unusual circumstances. One owner-driver friend of mine, from a country that will remain nameless, operated his 141 with a 111 badge on the grille. When I asked him why, he explained that the badge would account for the truck's sluggish performance when it was drastically overloaded.

Although other manufacturers have long since caught up with Scania, none has produced a model that evoked the same lasting enthusiasm as the Scania 141.

GEORGE BENNETT
September 2008

Introduction

Scania-Vabis L75

There comes a seminal moment in the history of every vehicle manufacturer when it produces a definitive product, a product that becomes synonymous with that company and which will represent its values and ideals for many generations, certainly in vehicle terms, long after it has been superseded by newer models.

Within the haulage industry these iconic vehicles will prompt steely-eyed transport people to state twenty years on: 'If they still made them I'd buy them today.' A slightly rose-tinted view maybe, and in these days of constantly changing legislation no longer an actual possibility; but nonetheless this type of reasoning is based on the great experiences and fond memories that can only be generated by an extremely good and competent product.

At the end of the 1950s Scania (strictly speaking Scania-Vabis as the two companies had merged in 1911) made significant steps towards producing their definitive product with the introduction of the L75 and its variants the LS and, slightly later, the LT75. These were ostensibly the traditional, tough Scania trucks for which the marque had become known featuring conventional bonnets, strong chassis and rugged running gear. However, the company's engineers had made some significant design changes over previous models, most of which were aimed at improving the driving environment of the new types.

A pioneering move, certainly for a European truck manufacturer, was the introduction of a unitary construction for the cab, wings and bonnet. Supplied by Be-Ge Karosserifabrik, a cab production company which would be acquired by Scania in 1966, the new assembly brought the benefits of a fully sealed bulkhead between the driver and the engine to lower in-cab noise and drastically reduce heat and fume induction. Further comfort was gained by mounting each corner of the structure on rubber blocks to absorb vibration and road shocks, a system that would also allow a 10 per cent flexing of the chassis without causing problems.

As the engine was now effectively cocooned by the new, one-piece wing and cab structure, a distinctive bonnet design was arrived at which raised the panel horizontally up and back on ingenious hinges. This feature was often referred to as the 'Alligator bonnet' as it mimicked that reptile when in the raised position. In another break from tradition the L75's bold, slated grille was flanked by neat, flush-fitting headlights mounted in the distinctive deep-section wings instead of the old pod-style items.

Inside there were changes, too, with instruments that were neatly clustered directly in front of the driver, rather than in the centre of the dashboard, and a new design of brake pedal which placed the floor-hinged item at the same height as the accelerator, making operation far more comfortable for the driver as well as improving safety. The braking system itself was revolutionary for the time with an advanced, dual circuit, air-operated system powered by a Bendix-Westinghouse compressor. The cab floor was flat and uninterrupted creating a welcome feeling of space and allowing easy cross-cab access. If the driver couldn't get comfortable on the new, adjustable seat there was also a sprung version available as an option and most models were fitted with power steering.

The motive power for the L75 was an all new in-line 6-cylinder engine designated the D10 which would go on to enjoy, through various developments, a career of over thirty years. In 1958 this 10.25 litre unit utilised direct injection to produce a healthy output of 165 bhp with a flexible working speed between 1200 and 2200 rpm. From 1961 the D10 would

South America became an important export market for Scania in the early 1950s when Vemag, a Brazilian import company, started selling the L71. This arrangement evolved first into an assembly operation and finally the opening of Scania's own extensive plant for the production and assembly of truck and bus chassis. Note the prominence of the sprocket logo in the days before Mercedes-Benz vetoed its use. (Scania AB)

also be available, fitted with an exhaust-driven turbocharger to become the DS10 ('S' for supercharged) which saw the output boosted to 205 bhp .

Although this marked Scania's first turbocharged engine to be used in a truck, the company (much like Swedish rival Volvo) had been using and developing the technology since the early 1950s which put them amongst the pioneers of the concept and helped create the edge which served them so well through the sixties and early seventies. Designed from the outset to handle forced induction, the D10 was to prove a very reliable unit in service and an ideal base for subsequent developments.

Power was transmitted to the rear axle by a robust five-speed gearbox, the G660, which featured synchromesh on all but first gear, and which marked another advance aimed at improving the driver's lot. As an option, the G660 could be coupled to an auxiliary unit to provide five further ratios selected via a second gear lever. The final drive options were double or single reduction types both of which featured an air-operated differential lock facility from 1959 onwards.

The model was available as a 4 x 2 (L75), a 6 x 2 with rear lift facility (LS75) and a 6 x 4 (LT75) with various wheelbases options from 3.4 to 5.0 m. 15,270 examples of all types were produced in the five years from 1958 to 1963. The L75 marked in fine style the start of an epic journey for the model which would include three rebirths by 1980; perhaps the most significant redesign being that of the all-important LB76, the next stage of the L75's incredible 22-year existence.

Scania-Vabis LB76
The Most Significant Ancestor

Scania had already made some concessions towards the standardisation of vehicle lengths in 1958 with the launch of the L76, a model which featured a shorter bonnet than previous designs to help reduce the BBC (Bumper to Back of Cab) dimension and allow legal coupling in a number of countries outside Sweden.

However, by 1963 the situation had moved on further with the world-wide introduction of the ISO container system which, along with the local laws appearing in many EEC countries to accommodate it, was having a profound effect on the shape and design of heavy trucks throughout the modern world. Scania realised that they needed to offer a forward control alternative in their heavy truck range if they were not to be left behind by their competitors. Although a significant step, this was not the company's first attempt at a forward control truck as they had produced an example up to thirty years earlier. However, the LB76 would be the turning point at which Scania's production of conventional trucks started to decline in favour of forward control.

Designated the LB76 (the 'B' relating to the model's nickname of 'Bulldog' after its cab shape) and first appearing at the 1963 Brussels Show, the new model was designed around the proven chassis and running gear of the L75 but with up-rated axle and suspension components at the front to take the weight of the new steel cab. The cab itself was of the fixed variety, something of a shame really, as Volvo already offered a tilting facility on their Raske model which used an early version of the F86 cab. However, it did feature a swing-out radiator which went some way towards improving serviceability; and access through the large cover in the cab, although not perfect, was at least acceptable.

Inherently strong, thanks to Sweden's stringent safety laws, the cab was of a clean, rounded design with good aerodynamic qualities. However, interior space was somewhat limited by the intrusion of the engine, an inevitable consequence of building a low-height cabover on a chassis originally designed for a conventional layout. From the start the cab was available in day or sleeper format, the latter fitted with a narrow rest-type bunk. To rationalise the design the sleeper variant was achieved by simply adding a section to the cab between the back panel and doors, a good cost-effective system that would also be used later for the LB110.

Under the new cab was a further developed version of the direct injection D and DS10 6-cylinder engine of the L75. With a marginally longer stroke the new unit's swept volume grew to 11 litres giving outputs of 190 bhp for the D11and a serious 220 bhp for the turbocharged DS11. This useful increase in power did much to establish Scania's position among Europe's top performers. During the life of the LB76 the engine was developed on two further occasions, in 1964 and 1967. With each revision the DS11's output grew by another 20 bhp resulting in a premium unit that produced 260 bhp at 2200 rpm and 710 lb/ft of torque at 1400 rpm.

Initially the gearbox on offer was the familiar, synchromesh, five-speed G660 of the L75 with the T2 again available as an option to gain the further five ratios. However, as the G660 had proved to be a weak point in the earlier range and, with engine outputs on the increase, Scania's engineers had been busy working on a replacement. The G600/1 became available in 1964 and replaced the former G660/T2 combination with a single and much tougher unit which also shaved some 50 kg from the chassis weight. The G600/1 would also form the basis of the forthcoming G670/1, a true ten-speed, range change type of box operated from a single lever fitted with an air-operated switch for the range change. Final drives were also strengthened at this point to cope with the extra power and were still offered as single or double reduction on 4 x 2 and 6 x 2 chassis both with air-operated differential locks.

Once again, a full dual-circuit air brake system was employed using a Bendix-

Westinghouse compressor with a two-reservoir layout. From 1966 onwards, the model also featured a modern-style, spring chamber, parking brake system. Designed for the UK, this was operated via a lever mounted on the dashboard. Power steering by ZF was once again standard throughout the range. Wheelbase options from 2.8 up to 5 m were available for the two-axle LB models while the LBS ranged from 3.5 to 4.6 m.

Drivers instantly took to the modern design of the cab, enjoying the comfort offered by the US-style counter sprung seats and the efficient heating and ventilation system. Power steering and synchromesh gears were a luxury and, because the performance was so strong, the trucks were easy and relaxing to drive. Reliability was fast becoming something to talk about too. More than any previous model the LB76 was helping to establish Scania's credentials throughout the all-important European market that was so necessary for the company's continued growth.

Jameson was among the first to operate the LB76 in Great Britain and would progress to the LB110 from 1968 on. Despite its fixed cab, the LB76, with its powerful engine, synchromesh gears and comfortable cab, was a revelation to UK drivers in 1966. Note the beautiful simplicity of the chassis hardware that was all neatly mirrored on the opposite side. It's probably best not to wonder quite what the Army are unloading in these drums. (Photo: West Pennine Trucks)

Notably from 1966 on, the LB76 was also made available in Great Britain, a particular corner of the EEC that was crucial to Scania's plans and one that had thus far been unavailable due to the crippling import duty imposed on foreign trucks. Over its five-year production life 22,717 examples of the LB76 took to the road, an impressive number for what was such a bold new design. With driver and operator acceptance high, the way was paved for the next major development in the LB story.

Scania LB110
A Lesson in Excellence

Scania launched the remarkable LB110 to a rapturous industry at the Amsterdam Show on 10 February 1968. Once again the underpinnings of the sturdy L75 were used as a base for the new truck. However, this latest redesign not only benefited from the ten years of careful development afforded to its predecessors but also rode in on the highly successful coat tails of the LB76 which, being forward control, had done much to modernise Scania's position in Europe. Scania's customers, both present and future, could not have been better primed for such an important new vehicle launch. Notably it was at this point that the VABIS name was dropped from the vehicles and, due to pressure from Mercedes-Benz, so too was the old 'sprocket' emblem which had adorned Scania's vehicles since the beginning of the century.

Undoubtedly the biggest single difference between the LB110 and the LB76 was the superb new cab which was complete with a tilting facility. Despite initial indifference to the idea of tilting cabs, Scania's engineers were given the green light to produce an example way back in 1964, which in fact was not long after the introduction of the LB76 when it was becoming increasingly obvious that tilt cabs were undoubtedly the way forward in Europe.

Much like its competitors, Scania looked initially at the products of the American manufacturers. Tilt cabs had been around in the USA since the late 1940s and had become very popular for maximising load length while staying legal. Also, much like its competitors, Scania's engineers soon realised that by applying some European finesse and engineering knowledge the basic concept could be dramatically improved. The resulting cab, actually designed by an Englishman, was one of remarkable good looks. High, wide and handsome, its square-jawed appearance was also fresh and modern, especially among its more rounded contemporaries, and somehow remained so for the next twelve years.

Manufactured from steel, the welded construction gave the necessary strength to pass Swedish impact and static weight laws and a 60 degree tilting angle gave good access to the engine and ancillaries. Tilting the cab was facilitated by an hydraulic pump system with a two-way valve which was reversible for lowering. The handle for the pump, when not in use, was neatly stowed in the right-hand floor beam of the cab, accessible when the front grille was raised, where it was secured with a spring clip to prevent it from rattling. Also behind the grille were daily check items such as the dipsticks for oil and power steering levels, fluid reservoirs for brakes and clutch, and also the cold-start system which was located beside the radiator.

Although it was considerably taller than the LB76, cab entry was easy with big doors that opened to 90 degrees; two well-placed steps and sturdy grab handles aided the ascent. Once inside, the driver found a level of luxury previously unseen in a truck and more like the standard of an executive saloon car of the day. As the cab was mounted so much higher, an engine cover no longer dominated the interior. This was now a low-height affair between the seats topped with a useful document tray and, due to the large glass area and the light blue/grey trim, the environment was spacious and light. Sprung seats with breathable vinyl covers for both driver and mate added greatly to the comfort, as did the dual heaters and the excellent ventilation which was facilitated by a multi-position roof hatch.

When specified, the sleeper version came with one or two foam-filled bunks, full width apart from a detent behind the passenger's seat and a wardrobe hanging area at one end. Stowage was generally good for the day with lockers in each seat base and under the lower bunk, plus tool holders on the rear bulkhead for the wheel-changing equipment.

The motive power for the range was the familiar D and DS11 engines. Still in their 1967 format these were proven from fitment in the last of the 76 range; they gave 260 bhp when turbocharged or 190 bhp for the normally aspirated version. The 1967 modifications to the DS11 also brought a lower, more manageable compression ratio of 15.5:1, a new camshaft, pistons and valve timing. At the same time a water-fed oil cooler was added which, combined with Scania's unique centrifugal oil filter system, further enhanced the engine's already legendary durability.

Although they had been further strengthened, gearboxes were the same as had previously been offered on the old 76 range with five- or ten-speed variations, the latter via an air-operated splitter option. This was unfortunate as there were definite shortcomings in the designs and problems were not uncommon in service. There was also a carry-over of the rather tall first-gear ratio which meant that pulling away on anything other than a billiard table type of surface was not only difficult to achieve smoothly but caused stresses to the rest of the running gear, especially prop-shafts.

However, all this was to be solved in 1971, with the introduction of the now familiar ten-speed range change synchromesh GR860 in which electrically selected planetary gears gave high or low ratios to the five main gears. The new unit was not only easier to operate but was also lighter, stronger and more compact than previous types and, crucially, it now contained a sensible first-gear ratio too. The introduction of the new gearbox was to coincide with the arrival of a new final drive unit, the R751, which combined first-stage reduction in the unit with secondary reduction in the hubs. These new components, combined with a change to stronger prop-shafts, fully addressed the old problems of weaknesses in the drive line and illustrated Scania's approach of constant development. Also in 1971 came stronger chassis frames (a move that was mirrored across almost the whole Scania range) which were engineered to handle all foreseeable developments in power and application and which had been developed in 1969 for the powerful LB140 range.

As with the LB76, there were 4 x 2 (LB110) and 6 x 2 (LBS110) versions available with a 6 x 4 (LBT110) variant added to the range in the summer of 1968, with wheelbases from 3.1 to 5 m and 3.1 to 4.6 m respectively.

Other minor changes that occurred over the vehicle's five-year life were the introduction of laminated windscreens, Bostrom seating, more prominent badges, larger mirrors and, most significantly in 1973, a slightly restyled front end which saw new, round headlights repositioned above the bumper and a new one-piece grille.

Over 23,000 examples of the LB110 were produced and the vast majority were supplied as 4 x 2 tractors which clearly illustrates the type's ideal specification and ability as a general haulage machine, a role it performed superbly.

B & W Motors Ltd (Bradburn and Wedge) delivered their first LB110s in the September of 1968 to Wiliams Bros. However, this early example was destined for the tanker fleet of MRTS based in Aldridge near Walsall. As such, it was equipped with the necessary Pet-Regs front-mounted exhaust system illustrated here. Interestingly it was also fitted with a spare wheel and carrier in the location usually reserved for one of the 44-gallon tanks which would typically be fitted to early models. *(Photos: West Pennine Trucks)*

The International Synthetic Rubber Co Ltd of Southampton were among the first UK customers for the new LB110 with this 1968-9, G-registered example recorded by the sales department of B & W Motors Ltd in this pre-delivery shot. Note the small and unusually mounted rear-view mirrors with which the LB110 was originally equipped and the remarkably discreet name and model badges, especially the barely perceptible 'Super' which denotes the fitment of the turbocharged DS11 engine with 260 bhp output in this example.

The Peterborough Show in 1968 and B & W Motors Ltd proudly displays Scania's new wares in the remarkably unglamorous style of the day which allowed the customers to be wowed by the product rather than the surroundings. The tilt cab of the new model was of course a crucial feature in 1968 and put Scania among the front-runners of the manufacturers that could offer this new and revolutionary facility. A generous tilt angle of 60 degrees made servicing easy and greatly facilitated engine and gearbox removal. The tilting operation itself was performed with the aid of an hydraulic pump mounted on the off-side chassis rail which contained a valve to reverse the process for lowering. *(Photo: West Pennine Trucks)*

Frigoscandia realised the potential for frozen foods in the late 1940s but, like Scania, enjoyed massive growth throughout Europe in the 1960s. Much of this growth was achievable because of the advances made by truck manufacturers which were supplying more powerful, more comfortable and more reliable vehicles. This stunning early LB110 drawbar combination was clearly brand new when photographed by Scania's publicity department and was the perfect tool for Frigoscandia's high-cube frozen loads. Note the ladder access up the side of the cab for the refrigeration unit and the insulated and chilled trailer. *(Photo: Scania AB)*

Sunter's, with great experience of heavy haulage and abnormal load work, were quick to spot the potential offered by the LB110, and PVN 468G was the first in a long line of Scanias to be honoured with the famous livery. This load, comprising a single concrete beam on a standard 40-foot flat, hardly rates as excessively heavy or indeed abnormal but would doubtless have given the DS11's 260 horses a decent enough work-out. Presumably the additional mirror on the nearside arm was for the use of the driver's mate as it is well beyond the driver's line of sight. Note the period, combined headboard/sheet rack. *(Photo: Tom Cook)*

HE Dobbs grew out of a small garage business based in Mill Hill, Enfield and was co-owned by Jack Redburn once he had sold his own concern, LT Redburn. Jack's son, Chris, went on to found his own haulage company, the concert and exhibition specialists, Redburn Transfer which also favoured Scanias for many years. The services of Dobbs' Scanias were often employed by the engineering company, Bristows, to move abnormal loads such as the two illustrated here. They also frequently transported imported nickel (used in the manufacture of turbine blades) from Africa for Johnson Matthews.

(Photo: Chris Redburn)

Despite having all the globe-trotting credentials of left-hand drive, a full factory sleeper and TIR plates, this well-travelled LB110 was photographed loading nowhere more exotic than Weymouth docks in the winter of 1972. Jameson (International Road Transport) based in Heywood, Lancashire was no stranger to Scania products, having been among the first operators to run LB76s with which they successfully established regular services to Paris, Antwerp and Le Havre. The subsequent LB110s continued to expand the company's routes throughout Europe. At the time of the photograph, OTB 271G was loaded with Channel Island tomatoes carried roped and sheeted! *(Photo: Adrian Cypher)*

By 1970 Scania's marketing people realised that they had been hiding the LB110's light under a bushel and dramatically increased the size of all the letters used for make and model identification. This magnificent LBS110 (6 x 2) operated by Unifos illustrates how the high visibility approach that was adopted made the SCANIA legend in particular far more prominent. Note also the subtle spray deflectors fitted to the cab corners below the indicators and larger mirrors mounted on the original-style arms. *(Photo: Scania AB)*

Scania's PR department made good use of Pentus Brown of Leighton Buzzard in the early days of the LB110 as they were the recipients of the first UK-registered example. JMJ 848G, pristine and brand new, would often feature in brochures and publicity material of the time, usually photographed around London's landmarks with a Boden three-axle trailer. In contrast MMJ 973H, looking particularly well worked, is pictured in May 1972 mixing it with the other traffic in London's Whitechapel district. With the fleet name 'Harbinger', presumably of doom for lesser machinery, this unit was, despite the lack of a 'Super' badge on the grille, equipped with the turbocharged DS11 engine and the proven synchromesh ten-speed range change gearbox. At the time synchromesh in heavy truck gearboxes was almost uniquely championed by Scania and Volvo and was a feature that did much to win over drivers. *(Photo: Joe Donaldson)*

Although it is difficult to make out the configuration of the chassis from this angle, it would be reasonable to assume, given the nature of the trailer and load, that this fine machine was an LBT110. Being a pre-1971 unit it would have featured the original 6 x 4 set-up offered which was carried over virtually unchanged from the original L75 and subsequent LB76. Post-1971 LBT chassis benefited from the vastly improved RP755 (later RP750) final drive, which offered hub reduction as a further option for really heavy operations. British Ropes was another company that graduated to the LB110 range following good experiences with the LB76. It is interesting to note, as this was a publicity shot for Scania, that the unit has had the badges removed.

(Photo: courtesy of Adrian Cypher)

Union Cartage ran twelve of these interesting 6 x 4 110s in their fleet. The official designation of these models would have been LBT110 with the T (standing for tandem) denoting a 3-axle chassis with a 6 x 4 layout; the 6 x 4 version became available six months after the initial launch of the LB110. London-based UCC originally ordered the big 6 x 4 machines as they believed, like many at the time, that the UK weight limit would be increased in line with the rest of Europe and, had that increase happened, the Scanias would have proved fine. However, the tandem axles caused extra drag and a hefty $1\frac{1}{2}$ ton penalty. This proved to be too uneconomical so eventually they were all converted back to a 4 x 2 configuration. AMP 657H is seen pulling hard on the A4 by the Ridgeway Café in 1972. Note the minor damage to the nearside corner and Jury-rigged indicator. *(Photo: Adrian Cypher)*

It was a lull in the timber industry that led Richard Green to borrow a few of the family firm's Scanias to undertake continental work in the mid 1970s. YDM 963H, a 4 x 2 unit with day cab, was the first to go abroad, piloted by driver Brian Dale, and is pictured at rest in Austria under clear blue skies. From that point on the Arthur Green fleet could be found gainfully employed on both round timber and TIR haulage with Scania LB110s and, later, LB111s and a sole LB141, proving themselves capable of tackling both tasks with equal competence. Note the more conventional mirrors and arms fitted to this example. *(Photo: Richard Green)*

This smart German-registered tanker combination illustrates the acceptance of the LB110 across Europe, even in countries which were traditionally very loyal to their own manufacturers and which imposed operating requirements to protect other interests, such as the railways. Note the long wheelbase of this 4 x 2 tractor and the unusual square surrounds to the additional spotlights. *(Scania AB)*

The appropriately registered FWW 110J was Arthur Green's first new Scania. An LB110 Super fitted with the turbocharged DS11, the unit is pictured fulfilling its intended role coupled to one of the company's log carriers in the days before Arthur's son, Richard, started running the units on continental work as well. Note the small mirrors and the fuel tank which has been repositioned behind the cab in an attempt to limit damage when off-road, although the exhaust was to remain vulnerable in its original position. *(Photo: Richard Green)*

This excellent 'schoolboy's' view over the wall of Swains' yard in the early 1970s was a chilling study for those British truck manufacturers then present. TUX 135J and RUX 887H performed so admirably that they were to have a profound effect on the buying policy of the previously all-British fleet. Following in Swains' tradition, the units were given the names of Crossways Poppy and Duchess respectively. The system for naming the fleet stemmed back to the family's horse stud which was called Crossways, with the latter name being that of one of the horses kept and reared there. Note that even with the up-to-date Seddons in the line-up, the Scanias are the only tilt cabs present.

It wasn't long before continental loads were undertaken with the new Scanias. The maiden voyage was a trip to an agricultural show in Spain with the Scania rep for the area, Dick Phillips, riding shotgun. Although not a factory sleeper, TUX 135J was fitted with a 3/4 bunk in place of the passenger seat which combined with the driver's seat to give a full bed. Driver John Clee preferred this arrangement to the less suitable fold-down version fitted in other LB110 day cabs and found that the storage space beneath was very useful.

(Photos: Swains of Stretton)

BRX 13J started life as a day cab unit on the Sayers fleet with the name, 'Wessex Clipper'. Although the LB110 was available with a factory sleeper from the start, there was a good deal of business to be done by the after-market conversion companies in the mid to late 1970s. This was partly due to the move away from drivers' digs but also because large numbers of day cabs were becoming available on the second-hand market. Conversions from the likes of Jennings or Locomotors were always easy to spot as they retained the day cab mud wing and would generally, as illustrated, feature a window in the added section of cab, unlike the factory option. *(Photo: Marcus Lester)*

Unfortunately anonymous in a plain orange livery, this early left-hand drive LB110 is nonetheless a good example of the neat factory sleeper option. Note the different front wings stretching back to the rear of the cab which were a key feature of the extended cab. The large doors concealed two well-placed steps which, being ahead of the front wheels, afforded easy access to the cab interior. Operating post-1983 at 38 tonnes, this example was probably a continental import operated by an owner-driver. MTW 262J has been fitted with a much later (post-1974) Scania badge and mirrors plus a Hatcher sun visor. *(Photo: David Wakefield)*

Although CTD 142J belonged to Merrick Transport and wore their smart burgundy and red livery, it was actually under long-term contract to Jameson Transport whose name is on the front of the roof and whose style of livery, though not the colours, and sign-writing was almost identical. Whether Jameson's had any influence over the purchasing of Merrick vehicles under contract to them is open to debate, but following the LB110 period the two companies' buying habits differed considerably with Jameson opting for Guy Big-Js whilst Merrick favoured Leyland Marathons. A return to British-built trucks was an unusual step at the time and is maybe indicative of this country's good initial response to the import threat. CTD 142J was photographed in Weymouth docks, a popular haunt of Jameson's own trucks. The year was 1972. *(Photo: Adrian Cypher)*

Even the mighty BRS could not ignore the national trend and started to succumb to foreign imports in the early 1970s. By the time this photograph was taken in the summer of 1972, both Volvo and Scania were much in evidence within its fleets. BAE 872J was a Bristol-based unit and is seen westbound on the A4 with a neatly roped and sheeted load. Shirt open and windows down - remember the days before air conditioning?

Not very old when photographed in September 1971, BHT 396J looks in fine fettle as it wends its way through Highbridge on the old A38. The excellent 11-litre engine was noted at the time for its power and frugality, the latter helped to a large degree by very efficient combustion. The crown of the alloy pistons was uniquely shaped to create a vortex which swirled the air and fuel into a near perfect mix for burning. BRS LB110s were all finished to a very similar specification right down to the two small rectangular spotlights below the bumper in most cases. However, this one does seem to have gained an additional off-side mirror and an unusual blank plate below the bumper too. *(Photos: Adrian Cypher)*

With their continental business thriving it wasn't long before sleeper cab LB110s started to appear on the Swain's fleet, making the drivers' lives even more comfortable during their long trips abroad. YUJ 45K was one of the very first to be bought and was soon out on continental work. By now the Shropshire-based trucks were making regular runs to Italy, France, Germany, Belgium, Holland and the former Yugoslavia,and the big side of the sleeper cab made an ideal canvass on which to advertise that fact. YUJ 45K bore the fleet name of Crossways Queen. As the bumper legend suggests, this unit also performed heavy haulage tasks, usually with a King low loader trailer. *(Photo: Swains of Stretton)*

UKON Trucking operated up to eighty trucks from its Harlow base between 1973 and 1999 with Scanias playing a huge part from the early days on. LPC 297K, an LB110 Super with left-hand drive, was chosen for this load destined for Italy in 1974. Although the Lancer Boss DD82 weighed in at a hefty 30 tons, the rented TIP trailer with four-in-line super singles mounted on two axles across the bed had a capacity of 40 tons (20 tons on each axle) and the DS11's 260 bhp was more than up to the job of moving the combination. Pieter Kroon, one of the partners in UKON Trucking, accompanied driver Roger Newbury on the trip. The initial crossing to Cherbourg was extremely rough and the extra chains applied by the crew tightened to such an extent that they had to be cut away when the boat docked. LPC 297K is seen as Roger threads the combination through a narrow village somewhere deep in the south of France. *(Photo: Pieter Kroon)*

John Gaze spent the two years 1980-82 piloting this veteran ten-year-old LB110 day cab for Harris Haulage. A lot of the work, as illustrated here, involved the movement of York trailer axles, much of which was for export to the Middle East. Harris Haulage ran around ten trucks including another LB110 and two LB80s; the company was later be taken over by Airflow Streamline which was part of the Ford group. An interesting feature of LNN 802K was the fitment of twin round headlights in place of the rectangular factory items which were not noted for their efficiency and which, rather harshly, were sometimes referred to as 'continental candles'. *(Photo: John Gaze)*

Nörsman International was a collaboration between Harold Woolgar and his neighbour who had expressed an interest in road haulage. Harold was, and still is, in insurance but divided his time between the two jobs. His wife, Win, also obtained an HGV licence and became a driver. RDH 518K was the first of many Scanias with LB111s, LB140s and LB141s following over the years. This smart unit was fitted with the latest style of full-size mirrors, large SCANIA lettering, corner wind deflectors and the round factory-option spotlights mounted in the bumper. Coupled to a wide-axle spread tilt, it made a real head turner. Note that the Scania lettering was moved down from its original position to allow the sign-writer to use the deep space at the top of the grille. *(Photo: Harold Woolgar)*

'Sheena Ann', in the smart livery of NMT, negotiates the roads around Aldgate in London on a hot summer's day in 1973. Almost side on, this view of the LB110 cab shows just how short it was in its BBC (Bumper to Back of Cab) dimension and with the 3.2 metre wheelbase presented a very compact unit. This made the day cab a popular choice with operators as they were very versatile in its coupling; even with a fixed fifth wheel, a wide variety of trailers could be accommodated. Hailing from NMT's Shotts base, this LB110 Super was number 6 in the fleet and was fitted with a chrome-plated bumper - well it was the early 1970s! *(Photo: Joe Donaldson)*

A smart and well-preserved 1973 LB110 Super with sleeper cab slips through Dover in the early 1980s. By this time, not only was Scania's backup throughout Europe amongst the best available, it was also firmly established in far-flung areas of the Middle East too. This was largely thanks to the local acceptance of older models such as the conventional L75 in these areas. With a good reputation for reliability, anyway, operators and drivers were therefore very confident in taking loads further and further afield making the LB110 something of a benchmark for TIR operators. Note the original-equipment Scania spotlights in the bumper. These were more commonly a feature of continental LB110s, which would often have two sets fitted. *(Photo: David Wakefield)*

Fifty years after Latimer Crow started his company it was running 200 trucks and had become one of the UK's leading specialists in tanker movements. Loads varied and could be anything from sulphuric acid to chocolate, the latter carried in highly specialised, temperature-controlled vacuum tankers built by Universal Bulk Handling (UBH). To minimise empty running, Crow established a network of tank cleaning facilities like this throughout the country. The purpose-built gantry delivered steam and detergent combinations directly into the tanks via rotating sprinkler heads, leaving trailers free of contaminants and ready for a back load. Like many others, Crow's business suffered in the economic downturn of the 1970s resulting in the closure of its northern base and the company becoming part of the Transport Development Group (TDG). However, fleet numbers remained in three figures and the specialised work was retained. Although fitted with day cabs, the Crow LB110s also undertook continental trips with liquid loads under TIR carnets. *(Photo: Scania AB)*

MAT Transport's simple but unmistakable livery, here freshly applied to a new LB110 Super, was perhaps one of the best known on UK roads. MAT (Machinery and Technical) was founded in London in 1926 by Swiss-born Arnold Kuenzler. In 1944-5 MAT was also established in Switzerland with Wilhelm Moser at the helm. Although MAT started out as a freight forwarder, it grew into a major transport concern establishing depots throughout Europe with its own trucks, trailers, containers and even specialised railway wagons for the transportation of cars. The two companies were independent but co-operated closely over the years until the demise of the UK division in 2006. MAT Switzerland remains a going concern led by Christoff Moser, the son of the founder. *(Photo: Scania AB)*

An impressionable young boy, probably the driver's son, looks on as this LB110 Super of Pigott Foundations is unloaded on a London building site in 1973. The piling machine load was just one of many operated by this company which specialised in bored pile foundations, a booming industry amongst the cleared bombsites of the capital at the time. Note that this was the time before health and safety and there is not a single safety helmet or yellow vest to be seen! *(Photo: Scania AB)*

Although a well worked eight-year-old when photographed at Taylor's Café on the A48 near Newport in 1980, DNG 158K looks to be in good shape. D and M Saunders' high-visibility livery and the fitment of an LB111 bumper with round headlights (actually auxiliary lamps on the later model) have done much to face-lift this 1972 LB110 unit. Note also that later-style SCANIA badge, mirrors and corner deflectors have been fitted. *(Photo: Marcus Lester)*

Shiny and new, BRS Swindon depot's Scania LB110 Super UFC 572K has an equally new tandem axle tilt. This unit was originally ordered to allow BRS to run back-to-back trials against a Scammell Crusader on the same continental journeys. BRS had been heavily involved with the design of the Crusader and on paper it measured up well against the big Swede. However, the fixed Motor Panels cab seriously impaired driver comfort and, despite the big swing-out radiator and large access hatch in the cab floor, servicing was significantly more difficult. However, the Crusader was as tough as old boots and found a place with many 'no frills' operators (including the Ministry of Defence) and struggled on in production until 1979 with examples still being registered from stock well into the 1980s. *(Photo: Adrian Cypher)*

From 1971 on, all LB110 models benefited from the stronger chassis that had been developed for the LB140, which had appeared a year after the LB110 in 1969. This was a substantial bonus to operators like Arthur Green whose trucks spent a lot of time off road in rutted conditions where chassis twisting was a constant threat. *(Photo: Richard Green)*

Astran, one of the UK's pioneers of the Middle East runs in the 1960s, derived its name from Asia Transport. Good experiences gained with the LB76 on those arduous journeys led to the purchase of LB110s and LB140s. Indeed, the Swedish machines performed so well that the fleet was to remain predominantly Scania throughout the 1970s, '80s and into the '90s. Astran's striking yellow-and-red livery became synonymous with the 'Gold Run' and, through books, television documentaries, magazine articles and advertisements, did great service to Scania in proving the capabilities of their products in the toughest of conditions. MUC 597L, an LHD LBS110 (S denoting a 6 x 2 chassis), is pictured at Dover awaiting a ferry for Zeebrugge with a 31-ton GEC transformer destined for Damman in Saudi Arabia. The arduous 4100 mile outward journey was completed in an impressive twenty-three days. *(Photo: Astran Cargo Services)*

Over ten years old, JOW 229L of Central Road Transport powers down the M4 on its return to base following a continental trip with this three-axle fridge trailer. Standard continental LB110s had been running at higher weights for many years before the UK increased its limit to 38 tonnes. At the new limit, UK LB110s with the DS11 engine gave around 7 bhp per tonne, which was good enough at that time for all but the most demanding of routes. *(Photo: Marcus Lester)*

When an owner-driver takes the plunge and purchases a brand-new vehicle it cannot be done on a whim. When EG Slade found himself in that position in 1973 he made a calculated decision and chose to buy this fine LB110 Super safe in the knowledge that the Swedish machines had built a good reputation for reliability since their introduction in 1968. Scania did not bill the range as a cheap alternative, quite the opposite in fact, but it did emphasise the point that the whole-life cost of their trucks would be considerably less and that residuals would be very much better. CHR 548L is seen in 1974 parked up on the Porte Marsh trading estate in Calne. At the time the unit was sub-contracted to Swindon-based EH Bradley & Sons, whose low-loader trailer is coupled behind, to move plant between their various gravel pit sites. *(Photo: Adrian Cypher)*

Cast concrete pipe sections make an impressive load for this tidy LB110 Super of D.J.Light in 1976. Jack Light's haulage business grew from a one-truck operation in the 1950s to a fleet of around forty by the mid 1970s. The bulk of work for the Holcombe-based trucks was derived from the local ARC depot from where pipe sections were dispatched to locations throughout the UK. No great fan of British trucks, Jack's buying philosophy was biased towards powerful foreign machines and from the late 1960s both Scania and Volvo were much favoured. *(Photo: Adrian Cypher)*

Proudly bearing its TIR credentials, this beautifully turned-out LB110 Super in the smart livery of Clifford Bassett was nearly new when photographed passing through Highbridge in the summer of 1974. Despite its youth, the unit was already a well-travelled continental veteran by this time, although the load in this case was UK-destined timber. *(Photo: Adrian Cypher)*

The collaboration between the American security firm, Brinks, and the MAT Transport organisation resulted in this most unusual LB110 variation conceived for the continental transportation of gold bullion. No doubt MAT, being very familiar with the Swedish marque, chose the LB110 as the base for the vehicle, while Brinks' contribution would have been the security measures which included bullet-proof glass and gun-ports for the officers inside. However, whilst the fitting of the hefty grille shutter would have protected the vulnerable radiator, the obvious weak spot of the design appears to be the unprotected tyres. Note that the air vents behind the grille, which fed the heaters and blowers, have been plated over to prevent the crew from being gassed. The Brinks-MAT organisation hit the headlines in 1983 when their Heathrow depot became the stage for Britain's biggest ever bullion robbery. *(Photo: Scania AB)*

Rigid LB110s were often to be found engaged in distribution work on the Continent, but rarely in the UK where such duties were generally performed by the smaller LB80/81, so this example represents something fairly unusual for the UK. However, SDM 458M may well have started out as a drawbar combination, especially as the grille badge suggest that this example was fitted with the turbocharged DS11 engine giving 260 bhp which would have been a little excessive for a solo rigid at the time. *(Photo: Marcus Lester)*

Ring of Rochester was one UK operator that did run LB110 drawbar combinations and circumstance would inevitably lead to their trucks running solo on occasions. However, ONT 220M is seen running in its full and impressive format here as it descends Jubilee Way into Dover docks in the mid 1980s. Ring's specialised removal service to the Continent required a powerful, comfortable, long-range truck with a super cube capability. The LB110 drawbar fitted the bill perfectly and was also flexible for unloading in tight spots as the trailer could be left at a location for unloading first while the truck was parked up out of the way, or vice versa. *(Photo: David Wakefield)*

The biggest change to the appearance of the LB110 came late in 1973 when the headlights were moved to a new position at the bottom of the grille. This new location afforded the powerful, round units far more protection than the original rectangular examples. At the same time the round auxiliary lights mounted in the bumper below were introduced as an option. OWS 30M was one of a number of Pollock vehicles which ran in the Scotrans livery during the 1970s. *(Photo: John Henderson)*

OFS 714M, owned by Vince Staunton, was another early example, registered on an M plate, to feature the post-1973 front-end revisions. Looking every inch like a true Middle East veteran, OFS 714M was also fitted with the optional 88-gallon capacity diesel tank. This item gave a useful increase over the normal 44- or 66-gallon type typically fitted to UK trucks and would often be paired, one either side of the chassis, to offer a massive range. However, the big tanks were hard to come by in the early days and even dealers found they had to side-step the system and import their own supplies via a source in Holland. Note that this example, a left hooker, has a grille modification and features later mirrors and SCANIA badge. *(Photo: Marcus Lester)*

A very tidy example of the revised front end captured by Scania's publicity department in 1974. Glasgow-based WS Unkles used this LB110 on fridge work to deliver loads of North Sea fish to customers on the Continent. The new grille of the revised LB110 had become a unitary item with the lower part of the original now incorporated in the main grille which opened upwards as one piece to provide access for daily check items such as the engine oil dipstick. *(Photo: Scania AB)*

For over thirty years the beautifully presented Pollock fleet was an all-British affair with ERFs, Atkinsons and AECs much in evidence. However, things were to change dramatically after the introduction of the first Scania in the early seventies. The newcomer proved well suited to Pollock's long trunk route work and although it was not a company to put all its eggs in one basket, LB110s soon became prolific in its fleet. Resplendent in the Pollock's turquoise livery with individual hand-painted tartan, Odin's Raven is seen here with a load of wire coils. With the exception of imports brought in from the Continent, the last new LB110s were registered in the UK in 1975, just prior to the launch of the LB111, and as such wore N plates. Note that HSC 189N also has the later style SCANIA badge which was probably retro-fitted after this item became available with the introduction of the LB111. *(John Henderson)*

Left-hand drive and a registration letter at least four years too late for an LB110 points to this fine machine being a European model imported into the UK around 1979. With the LB110's strong reputation for reliability and performance levels that were still more than adequate ten years after its introduction, especially at 32 tons, good, used examples became very sought after by owner drivers and small operators and, if running abroad, imported LHD versions were perfect. Note that YMX 86T has been fitted with Siamesed tanks on the offside taken from a DAF. *(Photo: David Wakefield)*

Although it was designed and developed in the Northern hemisphere, the LB110 actually performed very well when operating at the opposite end of the spectrum in areas such as the Middle East. This can partly be attributed to the water-fed oil cooler (new to the DS11 at the introduction of the LB110) which, combined with Scania's unique two-stage filtration, kept the life blood of the engine cool. Added to that, the system used no filters and was thus easily serviced in the field even by those of very basic mechanical knowledge and ability. Pictured whilst taking a break in the late 1970s, this well travelled left-hand drive example operated by Astran is a post-1973 example with the revised front-end styling. Note the minor damage to the lower grille. *(Photo: Anthony Pickhaver)*

LB140

Enter the V8 Express

Scania engineers started work on a new 'big power' engine as early as 1962 when it became clear that market forces were going to be demanding more and more power over the coming decades. With the opening up of all the continents, but particularly Europe, trucks were travelling ever-increasing distances with constantly growing payloads. Many countries, notably Germany, were demanding that trucks operating on their roads would have to meet specific power outputs measured in brake horse-power per tonne in the coming years. This was partly so as not to slow down general traffic which was also on the rise and partly, in some cases at least, in a vain attempt to protect the freight business enjoyed by the railways.

As Scania engineers had already calculated the potential for developing the six-cylinder DS10/11 over the coming years they knew they would need a different configuration to meet their projected output targets for the future. Ten years earlier, Scania had successfully developed a couple of reliable straight-eight designs, the D821 for use in a coach chassis and, slightly later, the D815 for use in rail buses. The D815 of 1953 went on to become turbocharged and has a legitimate claim to being the first series production turbo-diesel of which over 700 units were supplied.

However, there are inherent problems with cooling of the rearmost cylinders in the straight-eight configuration and, with the move away from conventional trucks, it was doubtful that such a design would fit satisfactorily under a forward control type of cab anyway. So, somewhat inevitably, the engineers settled on eight cylinders but in a compact 90 degree V layout of 14.2 litres capacity which, with a low height and short length, would easily fit beneath all current and projected forward control cabs. The character of the new engine was to be low revving to minimise stresses and increase component long life, with masses of torque available from low down to aid drivability and improve fuel consumption by reducing the gear changes required to a minimum - the 'let it lug' philosophy.

Early prototypes of the DS14, as the engine was to be designated, developing a whopping 350 bhp and around 900 lb/ft of torque at 1400 rpm, were being successfully road tested as early as 1966. Exquisitely engineered, the new engine featured an immensely strong block which had ample capacity for more power as and when it was required. It had a balanced five-bearing crank with bearings that were secured by horizontal as well as vertical bolts in a belt and braces fashion that added further strength to the block. The wet liners were supplied with triple sealing rings, and the unique parallel cooling system served all cylinders equally. The lightweight alloy pistons had separate cylinder heads that were restrained by eight bolts each to contain the high compression ratio of 15:1. There were chrome-plated valves with changeable steel heads, and lightweight, hollow push rods. The direct injection system featured a pump that was centrally mounted in between the cylinder banks to keep fuel lines short and maintain an even pressure with special five-hole injectors for effective atomisation in each of the cylinders. Finally, there was an efficient two-stage, exhaust-driven turbocharger.

Another feature to help reliability in the new engine was the use of close-meshed gears on the front of the engine to drive all ancillaries, with the exception of the alternator. This system eliminated the weak spot of v-belts and was substantially quieter in operation than the chain-drive alternative. The lubricating oil was fed, via Scania's now familiar centrifugal filter system, throughout the block entirely by means of internal passages, negating the use of any vunerable external pipes. it was even fed through the crankshaft and up the connecting rods to the inside of the pistons to aid cooling, a unique feature at the time.

The new engine was fitted into a modified LB110 chassis with increased beam height, flange width, and metal thickness throughout. As the engine was so compact it was possible to fit the standard cab from the LB110, designed by Lionel Sherrow, and was offered as a short day cab or full sleeper with one or two bunks. The new model range was to be known as the 140 with LB, LBS and LBT chassis configurations available, and, although outside the scope of this book, the range was available also as the conventional L, LS and LT140 from 1972 onwards.

When introduced at the Frankfurt show in 1969, the DS14 made the new model the most powerful truck available in Europe and by a considerable margin too. The myriad of positive magazine road tests that followed the launch soon dispelled the belief held by many a cautious operator that such a big engine was going to be hugely thirsty and actually showed that productivity, in many cases, could be better than that of lower-powered machines, especially when operating over tough routes. Operators were quick, too, to recognise the long life benefits provided by the low-revving, high-torque engine principle which gave all the transmission components, not to mention the driver, a far easier time.

The only chink in the armour of the new super truck was the gearbox. Carried over from the LB110, this unit was already troublesome in service and was dogged by a first-gear ratio that was far too high. Like LB110 customers at the time, those buying the new LB140 had to put up with this situation until 1971 when Scania addressed the problem by introducing the legendary GR860. This was a true range change ten-speed box, which was not only smaller, lighter and stronger than the outgoing G672, but also featured a sensible first-gear ratio. Also introduced to the LB140 in 1971 was an improved final drive, the R751. This hypoid type of unit had already proved itself in the LB110 and was both stronger and lighter than the original unit fitted to the V8-powered trucks. LBT (6 x 4) versions of both models were now fitted with an improved double reduction bogie, which featured planetary gears in the hubs to provide a second stage of reduction that improved tractability and considerably reduced the stresses imposed on drive shafts.

The cab of the LB140 models enjoyed the same revisions as that of the LB110 in 1973, which saw headlights repositioned in a new one-piece grille, and also the revised style of badges when the LB110 became the LB111 in 1975.

Utilising the same cab meant that the visual similarities between the LB110 and LB140 were such that identifying the new model was not easy at a glance. Obviously the LB140 did feature the numbers 140 in place of 110 on the grille and there was also a reasonably subtle V8 logo at the base of the doors, but in general the model was fairly anonymous whilst stationary. However, once on the move the problem of identification was instantly removed by the glorious rumble and bellow produced by the DS14 beneath the cab.

Nearly ten thousand examples of the forward control LB140 rolled off Scania's production line between 1969 and 1976, when the improved LB141 replaced the model. This was an impressive total for such a premium truck at time when having more power than you needed was in its infancy.

With proving grounds like Sweden's mighty timber forests, Scania couldn't fail to develop a top-class truck in the LB140. The 900 lb/ft of torque that was available at just 1400 rpm was, in 1969, quite awe inspiring and made light work of the typical timber loads of around 35 tonnes. Towering performance like this had not been witnessed before in a truck that was also capable of fast highway operation. From this point on all perceptions changed as to what one could expect of a premium-class truck. From the start the LB140 benefited from the larger lettering already introduced to the LB110, as can be seen on this magnificent LBS variant. *(Photo: Scania AB)*

The perfect example of fast highway operation would be those 'flying Dutch flower men' whose adoption of the LB140 from the very start was proof enough for anyone of the vehicle's on-highway ability. Despite the flat local terrain and loads that would cube out long before any significant weight was reached, this time critical environment was the natural place for the big Scania. To this day super premium vehicles are still used in this role. *(Photo: Scania AB)*

Ships of the desert, natural and man-made, take on the harsh conditions of the Middle East. If ever there was a job that was tailor-made for a truck like the LB140, it was the 'gold run' during the 1960s and '70s. Vast distances, mountain ranges, road-less deserts, extremes of temperature and service facilities that were few and far between all conspired against man and machine. However, the more powerful, reliable and comfortable a truck was, the better its chances were of survival, and success in this arena would reflect well on all the products of a manufacturer.
Astran, which was already loyal to Scania, were early UK customers for the LB140 with this fine LBT (6 x 4) drawbar outfit fitted with van bodies. Note the non-standard, cylindrical tank, possibly from a Scammell or similar, which has been fitted on the off-side and the small V8 logo on the lower section of the door which marked out the early DS14-powered LB140s. *(Photo: Astran Cargo Services)*

Astran's LBT140, JLL 685K (with a similar vehicle behind) carefully boards the ferry at Istanbul in the days before the Bosphorus Bridge crossing. The Turkish company, Enka Construction, working in collaboration with the Cleveland Bridge & Engineering Company Ltd of England and Germany's Hochtief AG, completed the bridge in just three and a half years. Designed by an Englishman, Gilbert Roberts, the monster construction was opened in October 1973 and is over 1500 metres long with a 39-metre wide, six-lane roadway suspended 64 metres above the Bosphorus sea. A recognised east-west crossing, it did much to speed up the journey times of the countless trucks streaming east.
(Photo: Astran Cargo Services)

W & M Wood ran an unusual daily service to and from Switzerland in the early 1970s for ROBA , who derived their name from the company's key sites in Rotterdam and Basel. LB110s were already on the Woods' fleet when the LB140 was launched and it did not take long before examples of the big DS14-powered variant were running in Woods' blue and gold livery over the company's European routes. Being early customers with interesting TIR work brought the fleet to the attention of Scania's PR department which would often feature Woods' vehicles, both LB110 and LB140, in company literature and advertisements during the early to mid 1970s. *(Photo: Scania AB)*

Another of Sweden's impressive 52-tonne capable combinations (the limit was 51.4 tonnes), this example is an LBS140 rigid with trailer designed for the transportation of cement. These enormous 24-metre combinations made up much of Sweden's heavy domestic traffic in preference over Europe's tractor and trailer system. Drawbar combinations running on six axles had access to around ninety per cent of Sweden's roads. *(Photo: Scania AB)*

This fine LB140 has all the flavour of a European motor with its simple, no thrills livery and tandem axle tilt in tow. However, NHT 889L harked from deepest, sleepy Surrey and, 'just as it says on the tin', was usually engaged on TIR shuttle runs to and from Lisbon, a duty that its 350 horses would most certainly have been capable of performing in an Express fashion. A regular stop-over point for the outfit was Swindon's County Ground as seen here in 1975. *(Photo: Adrian Cypher)*

With the DS14's 350 bhp output and a tough double-drive bogie, the LBT140 became a natural choice for those involved in plant movements and heavy haulage. Before the introduction of the LB140, specialist trucks would generally have undertaken this type of work, many of which would be built to order for a specific role. These vehicles would often struggle to perform any other duties due to their weight, gearing and dimensions. With the introduction of the big Scania, operators like Pigott Foundations found a flexible alternative which could be bought straight off the shelf and which could earn its keep at a multitude of tasks. As an own-account operator, Pigott Foundations used this splendid example for moving the biggest pieces of their plant. *(Photo: Scania AB)*

Northallerton's finest. This impressive prime mover was a permanently ballasted LBT140 that joined the Sunters fleet in 1973 to work alongside the company's versatile LB110 tractor units. With the strengthened LB110 chassis, the LB140 made a solid base for the heavily engineered bodywork and ballast box that was needed to contain the weight required to stop the double-drive bogie from spinning when pulling, or pushing, loads in excess of 100 tonnes. The big slab-sided cab must have looked great in Sunters' blue livery with yellow sign writing. *(Photo: Malcolm Slater)*

AF Cannatella's superb drawbar outfit takes a breather at Leigh Delamere services in 1982. Although fairly rare here, the LB140 was a popular choice for drawbar operation on the Continent with LBS (6 x 2), LBT (6 x 4) and even long wheelbase versions of the LB (4 x 2) being adapted for the role. Indeed, the 140 chassis had been designed from the outset to accommodate bodywork of various types with the minimum of modification and strengthening required. Van bodies later replaced the tilt bodywork seen here, at which time the livery was also changed to that of 'Hydrotech Systems Ltd'. Note the air filter, which is visible below the bumper on the right-hand side. This big cylindrical unit was a handed item that would be sighted on the left or right of the chassis depending on the driving position as the steering box occupied the opposite area at the front of the chassis. *(Photo: Marcus Lester)*

PHK 170M, was another, slightly later, LB140 of W & M Wood. Unlike JHX 602K *(page 48)*, this unit was supplied with right-hand drive and, being post-1973, also had the same grille and headlamp modifications that were applied to the face-lifted LB110. What is interesting to note is that this unit is also fitted with headlamp wipers. These were a standard fitment on the conventional L140 range in many countries but were offered only as an option in the UK for the LB series. *(Photo: Scania AB)*

A highly impressive addition to the Sayers fleet in 1974 was SRX 833M. The unit, christened Wessex Ranger, ended its days remarkably early with its last sighting being as an engine-less hulk parked up in the yard of Pat Duffy in 1981. A fifteen-year working life was not uncommon for an LB140 so at just seven years old, one can only speculate at this untimely end. Note that the spotlights fitted in the bumper are not the factory-fitted items. *(Photo: Adrian Cypher)*

Jack Light's fondness for high-powered foreign trucks was obvious to anyone visiting the company yard in Holcombe near Bath where a mix of tractor units from makers such as Volvo, Scania and DAF were always to be seen. PAM 108M was added to the fleet following sterling service from the LB110s that were being run. With an extra 90 bhp over the LB110s, Light's LB140s proved very useful on the staple work of concrete pipe haulage for ARC and posted excellent journey times with great reliability and remarkably good fuel consumption. Scanias were never the cheap option, and the LB140 carried a hefty premium over lesser machines in the early 1970s but, given his canny reputation for closing a good deal, it is unlikely that this unit joined Jack's fleet at the full list price! *(Photo: Adrian Cypher)*

Beautifuly presented in the finest Keedwell tradition, PWV 499M takes a break from its toils in the late 1970s. Graham Keedwell, like his brother Ray, was an early Scania convert and found that the grunt of the DS14 suited his low-loader operation very well. Neither brother has ever completely left the marque since. Note that a Hatcher sun visor is fitted, an item which really suited the LB cab, and that there is a space in the chassis as a second diesel tank was not fitted.

(Photo: Marcus Lester)

The big DS14 engine of the LB140 ran at a slow, for the time, maximum of 2300 rpm. When combined with the excellent oil lubrication system with the cyclone filter and even cylinder cooling arrangement this low-stressed approach gave the unit a superb reliability record. With proper servicing, many operators, even those travelling to the Middle East and beyond, never experienced a single failure from an engine. This fine colour shot features another of DJ Light's early LB140s, NMW 207M, making ready with another trailer load of ARC pipes.

(Photo: Adrian Cypher)

An anonymous LBS140 (6 x 2) drops into Dover with a period Merzario road/rail tilt in the early 1980s. The LBS chassis was a popular choice for continental and Middle East bound drivers and operators as, equipped with an air-lift facility, this layout safeguarded against axle overloading and could also provide extra traction in difficult conditions without the weight penalty or drag of a 6 x 4 bogie. This left-hand drive example features the revised style of badges, as introduced to the LB111 series in 1975, with silver letters and numbers mounted on a black background. *(Photo: David Wakefield)*

Although sales of such a super premium truck were very good for Scania in general, UK-registered LB140s were reasonably thin on the ground, especially when compared to the huge numbers of LB110s that were being sold in the UK. Given that most examples were put to work on continental and Middle East routes, it is not surprising that many, as here, were delivered as left hookers. However, offering a right-hand drive version gave Scania a distinct advantage over Volvo whose answer to the LB140, the F89 of 1970, was only available with left-hand drive. This tidy example, photographed southbound at the Rownhams services in the early 1980s, is fitted with the factory sun visor. This one-piece item was pressed out of lightweight aluminium and was susceptible to wind damage: hence the struts that were included to anchor the edges to the A-pillars. *(Photo: Paul Hart)*

A truly stunning example of an LB140, and at least six years old when photographed, enters Dover via Jubilee Way sporting a fresh Frans Maas livery. No doubt previously engaged on equally demanding work, GCE 311N, belonging to Anglo-Dutch Annika Transport of Kettering, proves the durability of the LB140 design. This longevity made second-hand examples highly sought-after particularly with owner-drivers and the smaller haulage companies that may not have been able to justify the cost of such a vehicle when new. Note the replacement mirrors and the model identification badges that have been swapped left to right. *(Photo: David Wakefield)*

PHO 99P came to Mansel Davies relatively late in life via an auction at Shepton Mallet in 1981. The six-year-old unit was formerly owned by Paul R Maggs and then Frank Parsloe. The latter ran it first in his own North Somerset Haulage livery and then, while pulling containers under contract to Whitwills, in a blue colour scheme. Heading down the M4 in 1982, the unit makes a purposeful looking combination running at 38 tonnes with its baulker behind. *(Photo: Marcus Lester)*

Just one number away from Mansel Davies's LB140 was PHO 100P belonging to Pat Duffy and seen here in Dock Road, Poole in the mid 1980s. Pat Duffy broke up all sorts of Scanias at his large yard in Southampton and was probably responsible for the chassis stretch that transformed this tidy example from a 4 x 2 unit into a six-wheeler for 38-tonne operations. *(Photo: Marcus Lester)*

Looking quite out of place parked up in a sleepy Marlborough High Street, this fine 1976 LB140 was nearly new when photographed and possibly was yet to receive a full livery and sign-writing. The LB chassis, for both 110 and 140 models, was a beautifully laid out and clean affair with a single battery box mounted above an air tank on each side directly behind the front wing, followed by the diesel tank of 44, 66 or 88 gallons capacity. Diesel tanks were often fitted in pairs, one either side, but on occasions the space on the nearside was given over to a spare wheel carrier. Like most of its contemporaries, the big LB cab was not immune from fouling its mirrors and side glass in bad weather. This unit is fitted with Scania's own-pattern wind deflectors to help relieve the problem. Note that the V8 door logo was dropped from the LB140 around this time. *(Photo: Adrian Cypher)*

NGB 959P was Graham Keedwell's replacement for PWV 49M *(page 53)*. By specifying the LBT (6 x 4) chassis, Graham was able to improve traction as well as increase payload potential whilst minimising the risk of axle overloading. The double-drive unit featured three air-controlled differential locks, one for each axle and one between the axles themselves to ensure traction even in the toughest conditions. Double-drive chassis were offered with either a heavy-duty, two-spring system for low loader and off-road applications or a more forgiving four-spring system for highway use. Once again, note the fine livery and sign writing that was applied to Keedwell's machines.
(Photo: Marcus Lester)

Another continental adventure beckons for this LB140 as it descends into Dover with a step-frame van trailer in tow. LYH 282P was at least seven years old at the time but with its fresh and tidy paint would easily have passed for a vehicle half its age and no doubt performed like it too. As Lionel Sherrow's cab design was free from any particular styling details or gimmicks it possessed a timeless appeal which prevented it from dating for many years. In fact, it still looked contemporary when replaced by the modern Series 2 in 1981, some thirteen years after its introduction.
(Photo: David Wakefield)

Astran's old base in Addington, Kent was the start point of many a Middle Eastern adventure for the company's growing fleet of LB Scanias during the 1970s. KVX 859P, a very late 4 x 2 tractor unit registered in 1975 just before the introduction of the even more powerful LB141, is seen in fresh livery posed on the old garage forecourt in front of the large London Road depot. *(Photo: Anthony Pickhaver)*

Maybe 350 bhp is a little excessive for a shallow-bodied Tarmac-laying tipper but this well equipped LBS140 was also the prime mover for a 24-metre combination that allowed one truck to handle three bodies and up to 35 tonnes of payload. The aluminium tippers were hook loaded from either end of the semi trailer and then emptied by twin rams. Note that this 6 x 2 chassis is fitted with hub reduction to lower stresses in the driveline during the low-speed operation of laying the road and to assist getting the 52-tonne outfit moving. *(Photo: Scania AB)*

Big boys' toys! Scania's publicity department knew all too well that pictures could speak a thousand words and were constantly capturing impressive images like this to illustrate the abilities of the trucks. The LB140, seen here in LBT (6 x 4) guise, was a natural choice for those involved in abnormal loads and heavy haulage. In addition to the truck's power and traction, the new strengthened chassis (a further developed version of that used for the original LB110) made an ideal platform for this type of work and was able to absorb the punishment of operating on less than perfect terrain such as building sites. This ungainly load appears to be being carried somewhat precariously on bolsters, one on the swan-neck and one on the rear deck, of this standard low-loader trailer. *(Photo: Scania AB)*

Love it or hate it, the customising fairies did enjoy a degree of success when they visited the yard of Carmans Transport to turn MBE 914R into a Kenworth facsimile, a deception that extended to the removal of the third wiper and provision of a pseudo split screen. With the replacement LB141 being launched in 1976, R-registered examples of the LB140 like this were something of a rarity and probably represented cancelled orders. Note that even though the grille has been partly panelled over to look like an American truck, a series of small holes has been drilled through to allow the original cab vents behind to function properly. *(Photo: David Wakefield)*

A 1-100 ton towing capacity is the proud claim of this handsome wrecker operated by Billington Bros. While it may have looked ridiculous with a Mini on the business end of the Holmes 750 lifting gear, a fully laden artic would have been a different matter. Although wearing a Q-plate, the age of this example can be narrowed down to the last few years of LB140 production by the post-1973 revised styling. Note the chequer plate repair along the rust-prone seam above the mudguard and the sturdy three-pin towing hitch that could point to a past life as a heavy haulage machine. *(Photo: Malcolm Slater)*

Quite how an Irish-registered truck came to be based in Devon is not known to the author but EIB 6373 of Mahony Transport (a curiously Irish sounding name) makes an impressive sight as it resides by the quayside in Bideford in 1989. A well-worked LBS140 engaged on round timber haulage, the unit had obviously ventured off road recently leaving fresh mud on the underside of the air tanks. The 6 x 2 LBS chassis was a popular choice for forestry work and provided excellent traction with high axle loads and good on-road performance.

(Photo: Marcus Lester)

Having said that R-registered LB140s were fairly rare, here we have an even later example running on a 1978 S plate. Old dealer stock and cancelled orders would inevitably lead to this occurrence but it was also a fairly common practice for operators to buy new lorries and leave them unfregistered for a further six months or so. This deception would only become glaringly obvious when models were superseded by newer examples during the gap, as was the case with the LB140/1.

(Photo: David Wakefield)

Scania LB111
The Definitive Product?

By 1974 the LB110 had been in production for six years and over 23,000 examples had entered service throughout the world. In addition to that figure, Scania had also produced over 30,000 examples of the conventional L110, which shared the same major components as the LB range. With such large numbers in operation, Scania engineers were privy to a wealth of information and experience gathered from all manner of environments. As a result, the range benefited from an intensive development programme that, although led by the factory, was often heavily influenced by the customer. The improvements resulted in an exceptional range of trucks with outstanding ability across a broad range of applications. Although Scania was still a relative newcomer on the world stage, the haulage industry was impressed. Despite some significant changes during those six years (most notably the new GR860 gearbox of 1971) the LB110 designation was retained for the model and by doing so Scania prevented the range from becoming a blot in the company's history; nobody now recalls the LB110 as a duff motor that wouldn't pull away easily or which broke its prop-shafts with monotonous regularity. Instead, by sticking with the model and constantly improving it, Scania created a legend, a legend that by 1974 was ready to pass on its crown.

Having proved the merits of the low-speed engine principle so decisively with the DS14-powered LB140 of 1969, Scania now wanted to promote the benefits of this philosophy to the entire haulage industry and not just those operating in the top echelon. With this development in mind, Scania engineers embarked on a significant re-development of the D and DS11 engines with the decision that, as the LB110's work was now done, it should be re-launched in a new model. The 'New Programme Scania' initiative was launched at the Earls Court Show in September 1974 - enter the LB111. The purpose of the 'New Programme Scania' campaign was to educate drivers and operators on how to use the new products which worked at their economical best within a small designated band of revs which started around the 1300 rpm mark. It wasn't long before rev counters were marked with the now familiar green sector and Scania literature adopted the phrase, 'drive on green' to get the message across.

Although the capacity and stroke were to remain the same as that of the outgoing engines the power output was increased by around 10 per cent through the adoption of a new turbocharger and a careful re-design of the induction and exhaust systems. This meant a healthy 280 bhp for the blown DS version which placed the LB111 nicely towards the top end of the premium-class tractors then available and, maybe more significantly, 40 bhp ahead of Volvo's standard F88. The changes made to the valves, ports and manifolds also helped to lower noise levels and reduce emissions - which proved useful experience for Scania with the tightening of legislation in those areas looming. The low-speed characteristics of the new engine meant that there was now 804 lb/ft of torque available at just 1300 rpm which, combined with high ratios for the final drive, helped drivers to stay in the economy zone while still making very good progress. Scania demonstrated that, if driven correctly, the reduction in the number of gear changes required could mean an improvement in fuel consumption of around 5 per cent over the outgoing LB110. Other engine modifications included a larger oil pump and increased capacity for the cooling system, which now featured an expansion tank and a low-level device with a warning light on the dashboard. As an optional extra, buyers could specify a thermostatically controlled fan which allowed the engine to reach its optimum operating temperature faster than the standard fixed item, further improving economy and also reducing noise levels.

A standard single dry-plate clutch was used to transfer power to the gearbox with a double dry-plate option made available soon after the launch for those operating at higher than normal weights or in scenarios where constant gear changing was required – such as urban multi-drops. The now familiar GR860 ten-speed gearbox came unaltered from service in the LB110. Because of the synchromesh, gear changes were not the quickest in the world but the unit was now well-proven and was much liked by drivers. The basic five-speed gearbox (G760) without the range change was still available for a time for those who really wanted it, and later the GRE860, equipped with a power take-off facility, could be specified but only for use with the turbocharged DS11.

Final-drive options were as before with the choice of two single reduction units for normal use. The standard axle (R651) featured normal gear teeth of the Gleason type and was available with three different ratios. Alternatively there was a Hypoid option (R751) which, due to the larger contact area of the helical gear teeth, offered increased reliability, and was available with two different ratios. For extra-heavy work or difficult terrain there was the RP750 which was a single reduction axle combined with hub reduction. This was a tough unit that utilised four cylindrical planetary gears in each hub to reduce loadings through the rest of the drive-train. Two gear ratios were offered for the axle and, as the hubs worked at just 1.39:1, the thermal loadings were low. An air-operated differential lock was a standard feature of all axles and the 6 x 4 (LBT) chassis was equipped with three, one for each axle and one between the two.

Chassis frames were the improved version introduced to the LB110 (following development for the LB140) and featured U-section side members of cold-pressed steel with a constant depth of 270 mm and a uniform top surface to ease the mounting of bodywork etc. The layout of the chassis hardware such as battery boxes, air tanks and diesel tanks remained as before.

At launch the LB111 was available with either the short-day cab (type HB11) or the sleeper version (type HB13) with the latter offering one or two bunks and such luxuries as a roof hatch, curtains and reading lights. From outside there was little change apart from a bold new SCANIA badge below the windscreen and model identification which was now positioned at the bottom right of the grille. Inside, however, were new seats, heated and weight-adjustable for the driver. There were new fabrics in warm shades of brown and the dashboard now had a matt finish to reduce reflection and glare – all of which was aimed at making the driver's life more comfortable. Significantly, from January 1975 Scania also offered a third cab option - the HB13 deluxe. With a level of luxury not seen before, the new interior was developed to meet the demands of those employed on super-long distance haulage, specifically to the Middle East, which meant drivers living in their cabs for months at a time. Extra sound insulation, a fitted carpet for the floor and engine cover, electrically adjustable mirrors and a removable wardrobe did much to further enhance the already strong driver appeal of the big cab.

During the LB111's seven years in production, developments beyond detail changes were few, perhaps because the package was so good from the outset. However, significantly, in 1977 a more powerful version of the DS11 (the DS11-02) was made available which, mostly via changes to the injection pump, saw the LB111's output peak at 305 bhp. To complement the up-rated engines, gearboxes were strengthened and re-designated as G770, GS770, GR870 and a new 'super duty' hub reduction rear axle (RP830) was introduced. In 1978 certain models including the LBS111 (6 x 2) were made available with the same air suspension system that had been available on 4 x 2 models since the LB111's launch.

In addition to the usual LB, LBS and LBT models, of which nearly 30,000 were produced, Scania also built small numbers of a twin-steered eight-wheel rigid chassis known as the LBFS between 1978 and 1980. This variant saw service mainly in the home market, often as a prime mover for drawbar outfits. However, far more significant was the LK variant that featured a front axle located 415 mm ahead of its normal position. Produced between 1976 and 1981 the LK, much like Volvo's G variant of the F88 and F89, was developed to beat axle load restrictions in certain markets such as Australia and most notably Peru where 3,833 examples of the model were assembled from kits.

If the LB110 had been Scania's invading force, sweeping all before it and establishing a secure bridgehead, then the LB111 surely represented an occupying army.

Although the first LB111s did start to appear on N registration number plates they became far more numerous after the new P plates of 1975 were introduced. This could be down to the usual clearing of old-stock LB110s at good prices combined with the reticence of operators to adopt the new model when the old one was so good. Whatever the reason, it did not take long for the LB111 to assert itself in the marketplace. This example was one of many LB111s operated by ROBA, alongside LB140s and LB141s, from their Harlow base and is seen here close to completing another of the company's regular Swiss runs as it approaches its tenth year in service.

(Photo: David Wakefield)

In the author's humble opinion, Scania launched their definitive product in 1974. Although lacking the ultimate glamour of the V8 models, the LB111 was very much all things to all men, a tough, powerful truck with strong performance that could be put to work on a multitude of tasks. It would perform all of them reliably and economically while cosseting the driver in a safe and comfortable environment. This early example belonging to R Carter and Sons was flying the flag for Scantruck along with examples from Brain Haulage, Pan Express, Russel Davies and Seawheel at a truck show in 1975.

(Photo: Anthony Pickhaver)

James McBride's very smart LB111 'Forward Caledonian' waits at the Port of Leith during the long, hot summer of 1976. McBride's has since moved into the specialised area of glass recycling but originally it ran an interesting mixed fleet employed on UK general haulage. MFS 736P was fleet number three. Note the early-style badging of this LB111. *(Photo: John Henderson)*

ROO 678R, looking really purposeful with ladder rack, air-con unit and other Middle East accoutrements, sweeps out of Dover's East Dock in an impressive fashion. Essex International's forays into the Middle East were legend. For many operators undertaking this type of work a Scania was the weapon of choice, with the LB111 providing an excellent compromise over its bigger brother the LB141. Scania's support system was well established throughout the route and its products were well understood and highly regarded by the locals. Indeed the LB111's mechanicals had been operating in countries such as Iraq since the introduction of the L75 way back in 1958. Note how relatively clean this example is for a returning Gold Runner. *(Photo: Adrian Cypher)*

A nicely presented LB111 operated by SJ Barch seen running at 32 tons with this tandem axle curtainsider illustrates well the general haulage capability of the LB111. The model was a true jack-of-all-trades with a perfect balance of power and economy, which made it ideal for a vast number of different operations. Added to this, it also had a strong 'premium' image recognised by hauliers and, more crucially, their customers alike. *(Photo: Marcus Lester)*

Although not in its first flush of youth, this Essex-registered LB111 of Cambridge-based RE Harradine looks distinctly purposeful as it plies its trade down the M4 in the late 1980s. Although not running at its 38-tonne capacity, this combination appears to be carrying (from the author's experience in this field) a load of brewery barrel or bottle pallets, possibly destined for the cider manufacturers of Somerset. *(Photo: Marcus Lester)*

Looking resplendent in EST's vivid pink (actually Elizabethan purple) and yellow colours, SOO 821R heads up the M2 from Dover in 1984 while working on the Van Halen tour of that year (1984 was also the title of the band's album at the time). Although EST were starting to move over to DAF by this time, SOO 821R was bought while the fleet still predominantly comprised Volvos and was the choice of company director, Del Roll, as his own personal truck. The unit was one of a batch that had been based and operated in Saudi Arabia by Brain Haulage, hence the Essex registration. Similar to the TIR carnet system, the ATA plate on the grille represents a temporary import/export document to cover loads of this nature and also allows Sunday running on the continent. Although they seldom display the plate these days, EST trucks still operate under the ATA system and paperwork. Note the 'Super' badge, taken from an LB110, which has been added to the grille. *(Photo: David Wakefield)*

This is how Astran vehicles appeared for much of their time: dirty, travelled stained and sun baked. 'Desert Tramp', with the Essex registration SOO 816R, was a left-hand drive unit and a stalwart of the fleet which went on to amass an impressive Middle East record spanning many years. Note the rather unusual British-registered White Freightliner which is sharing the desert compound in this photograph. *(Photo: Anthony Pickhaver)*

EG Slade made the move to Scania in 1973 with the purchase of a brand-new LB110 Super with day cab. Having been impressed with that vehicle, the Wiltshire-based owner-driver continued his association with the marque with this smart 1978 LB111. Ted ticked the HB13 option box this time, so as well as benefiting from all the LB111 improvements, FCG 705S was also equipped with a full factory sleeper. The unit is pictured on the Porte Marsh Trading Estate with Ted's usual Crane Fruehauf flat behind. *(Photo: Adrian Cypher)*

Although the headboard tells us that this LB111 was employed by Steads, the familiar livery, bumper destination legend and the just legible 'Crossways' fleet name at the base of the grille marks this unit out as ex-Swains of Stretton. Doubtless the veteran of many thousands of European miles for its former owners, VAW 790S was still putting in a good day's work at 38 tonnes for Steads in the late 1980s delivering building materials with this self-loading flatbed trailer. Note the three-cylinder donkey engine for powering the crane. *(Photo: Marcus Lester)*

A decent coat of road film covers the chassis and two-tone livery of this well-worked LB111 of Freightline Transport. The LB111 made up the backbone of this Bristol-based fleet for many years from the late 1970s on. UK general haulage and continental traffic was undertaken, often with Crane Fruehauf box vans like this one, as well as produce in and out of the London markets under GLC excemption. *(Photo: Marcus Lester)*

WYC 257S was another of Freightline's LB111s. Company owner Tom Williams was well known for looking after the welfare of his drivers. He not only provided top-quality vehicles for them to drive but he also let them take their trucks home. This one was piloted by Tom Byrne and was photographed near his house in Bristol. It's a shame that amazing mixed loads like this are so often hidden away behind curtains these days. *(Photo: Marcus Lester)*

An anonymous left-hand drive LB111 drops into Dover with an empty, or part-loaded tri-axle fridge in the mid 1980s. Note the air conditioning condenser unit mounted on the roof hatch of this example. The factory-fitted item was neatly plumbed through the roof hatch with the controls and outlets housed on the underside. Scania recommended that the unit be run up for five minutes every week and that the intake be kept clear of obstruction – whether or not paint over spray affected its performance is not known. *(Photo: David Wakefield)*

The lost art of roping and sheeting expertly demonstrated by the driver of this smart Ken Oak LB111 parked up at Poole in the early 1980s. The wide panel at the top of the grille, which carried the SCANIA badge, lent itself to being picked out in a different colour which could then be run around the cab. Many operators took advantage of this and incorporated it into their livery. *(Photo: Paul Hart)*

Italscandia was established in 1974 as the concession for Scania in Italy. This coincided nicely with the introduction of the LB111 and the model proved to be very popular. Indeed, with the later LB141, Armando Rangoni, an ex-Fiat dealer, was able to grow Scania's market position to that of number two behind the domestic giant Iveco. Many of the LB111s operated in Italy were fitted with right-hand drive to assist the driver's view of the road edge in the mountains; Trilex wheels were de rigueur too in most cases. Note the over-spilling load of this fine example operated by Eurotrans. *(Photo: Scania AB)*

Was it the truck or the driver that was named Elizabeth? A happy female pilot beams for the camera, clearly proud of her smart LB111 of EIC Roadliners. The TIR tilt and pallet of plonk suggests continental trips were the order of the day for this intrepid combination. Note the large chassis locker, not believed to be a factory option, mounted on the off-side of this 4 x 2 example. *(Photo: Anthony Pickhaver)*

John G Russell's purple-liveried fleet was well established by the time this LB111 was bought in 1979. Originally the fleet had a heavy bias towards Volvo (usually short-day cab F88s), but Scanias started to enter it in the late 1970s. Although the modus operandi of the company had not changed, with trucks still trunking down from Scotland overnight to be tipped throughout the south by fresh drivers during the day, there was a definite shift away from day cabs around this time with full sleepers like this becoming normal spec for Russell trucks. The simple but effective livery remains the same to the present day and is still more often than not applied to Scania products. DGD 903T is pictured in Cavendish Square, Swindon which was a popular spot for Russell drivers to change over. Note the big 88-gallon tank, mirrored on the opposite side, which easily provided a round-trip capability without refuelling. *(Photo: Adrian Cypher)*

Barely requiring an introduction, Cadwallader's trucks with their simple red and blue livery were immediately recognisable and a familiar sight all over Europe. Despite a huge allegiance to Volvo (the fleet numbered over eighty examples at its peak) Cadwallader also ran a number of LB Scanias on its intense TIR work. This 1979 example was three years old when photographed over-nighting at the Swindon County Ground and looks to be in fine condition. *(Photo: Adrian Cypher)*

TE Jones's smart livery, although a little faded here, looks good on this well-worked LB111. The LB's steel bumper would often pick up stone chips which, if left unchecked, would soon fester into rust bubbles giving this pock-marked appearance. However, on the up side, this was indicative of a hard-worked truck that spent little time off the road. The same company's MAN parked behind seems far fresher in this respect. *(Photo: Marcus Lester)*

A crisp winter's morning delivery for this exceptionally smart LB111 of Lloyds of Ludlow. This style of traditional livery really suited the square cab shape of the big Scania, and Lloyds were not alone in repositioning the Scania badge to allow large lettering to be used below the windscreen. Note the extended Trombone trailer being used for this delivery of steel. *(Photo: Scania AB)*

Recording a nice link with the past is this Scania PR shot from 1978 featuring one of TH Brown's shiny new LB111s unloading paper reels at Hull docks along with one of the companies original and hard worked LB76s. Although a revelation to drivers back in 1963, the LB76 cab was quite crude in comparison to the LB111. Note how well Brown's simple livery suited the LB111. *(Photo: Scania AB)*

By 1978 Swains had enjoyed ten years' experience of LB110/11s and were in the full swing of Scania ownership. Having had no major problems or failings, and with driver acceptance very high, the company had no reason to look at other makes. Whilst still pounding the TIR routes of Europe there was still plenty of opportunity for domestic loads to be undertaken too as here with this roped and sheeted flat. Note the repositioning of the badges to accommodate the sign-writing and fleet name of the unit.

(Photo: Scania AB)

The LB111 was not a common choice for car transporter operators, most opting for far more humble trucks. However, with a nice flat roof and no air stack to reposition, the LB111 was nonetheless a good choice for the role as is demonstrated here by this Essex registered example in the employ of Hillsons Garages of Sussex. Note the large fuel tank which, although similar in appearance, is not an original Scania item. *(Photo: Marcus Lester)*

As the LB111 inherited the strong chassis developed for the LB140 it provided an ideal mounting point for a Hiab or similar hydraulic crane and as the back of the cab was so uncluttered there was no need for modifications. This was a boon to own-account operators such as this who, by the nature of their work, would often be delivering their equipment to building sites at the very start of operations before any other lifting gear was on site. Note the position of the second bunk. *(Photo: Scania AB)*

The once familiar livery of the Aston Clinton Haulage Company adorns this 1979 LB111 seen coming through Dover with this wonderful period twin-axle tilt. Generally speaking, the big Scania cab withstood corrosion reasonably well but high-speed motorway miles could cause front panels and bumpers to become peppered quickly. However, far more serious were the box sections and channels beneath the cab which would harbour moisture and cause the cab to rot from the inside out; this would often go unnoticed until it started to break through on the seams along the top of the front wings. Other common areas to suffer were the bottom of the windscreen surround and the gutters, but even these would not usually be a problem until a truck was well into its useful life. *(Photo: David Wakefield)*

Another well-known haulier to run LB111s was Dibb & Ross. The empty spotlight position in the bumper of WEO 222T gives a good view of the large air filter housing that was mounted at the front nearside of the chassis. This was a handed item that would swap places with the steering box on the other side of the chassis depending on the truck being left- or right-hand drive. The filter itself was a dry type and changing and cleaning were done by removing the bottom of the canister which was retained by four over-centre catches. The system incorporated a pressure sensor connected to a warning light on the dashboard which would glow at high engine speeds if a blockage occurred. Note the new, or recently refurbished, box van trailer. *(Photo: David Wakefield)*

Despite being a premium tractor with a premium price tag to match, the LB111 found favour with many own-account operators such as Lyons Tetley. Fleet managers in companies like this soon realised the benefits of performance and reliability offered by the LB111 and that the model enjoyed greater resale values than lesser products. EJN 514T is pictured pulling steadily up the A68 out of Dalkeith in 1982. Note the 1980s 'barn door' style air deflector.
Photo: John Henderson)

This LB111 of Continental Freeze Ltd was five years old when photographed at the company's Broxburn depot in 1983 but was originally registered way down south in Essex in 1979. All LB111 bumpers had the provision for one set of spotlights when they left the factory but, if they were not fitted, blanking covers like these were available to close the aperture.
(Photo: John Henderson)

Well, the livery is familiar but there's something different about the name and sign writing. XUX 927T was one of a small number of tractors operated out of Cherbourg by Swains of Stretton. The UK-registered, right-hand drive LB111 worked alongside French-registered Saviems and Berliets. Note that the unit still received a Crossways name but in this case it was Yvette. *(Photo: Swains of Stretton)*

Three of a kind. Variations on a theme for these LB111s of CGM, or Compagnie Générale Maritime, with each unit finished in a slightly different livery representing three branches of the same extensive French shipping company. Although all under the same parent company, these three units were all supplied and registered by different dealerships. *(Photo: Anthony Pickhaver)*

Bedale Transport of Patrick Brompton were big users of the LB111. Bedale was part of the John Lyon group and the smart red-and-white liveried tipper fleet was a common sight across the border in Scotland during the 1970s and '80s. Previous LB111s ran in the same livery but were sign-written with John Lyon Ltd in the white stripe and in the headboard. *(Photo: John Henderson)*

This well-worked LB111 of Hudsons started life in 1979 as a 4 x 2 unit but was given this tag axle conversion in later life (unofficially making it an LBS) to take advantage of the 38 tonne limit. Although running nowhere near that sort of weight, this low loader combination with hopper load makes an impressive sight as it heads down the M4. However, the TIR plate suggests this was outside the truck's usual line of work. *(Photo: Marcus Lester)*

McGawn Brothers' superb LB111 OHS 322V bowls steadily down the M4 many miles from its Maybole base in Ayr in the 1980s. David and James McGawn started their haulage business in 1968 with an AEC Mandator tractor unit with which they hauled timber for a local sawmill. However, an LB80 bought new in 1971 was to prove significant to the brothers' purchasing policy from then on. With the exception of two interlopers, neither of which was kept for more than one year, the fleet has been made up of Scanias ever since. The company ran five LB80s, six LB110s and six LB111s. OHS 322V was the last LB Scania operated by the company and worked on until 1992. Today the eleven-strong fleet totally comprises veteran 3-Series Scanias which, despite their age, are kept and maintained to the very highest standard. A regular sight at truck shows, the company frequently wins trophies for its vehicles. *(Photo: Marcus Lester)*

CAE 531V started its working career in 1979 with Western Transport. The day-cab tractor was used to fulfil an intensive day and night trunk service for Boots the Chemist. Parkers were based in the Yeovil area and used the unit to head up this 38-tonne bulker. Note the LB110 bumper, complete with the original rectangular headlights, which has been fitted. *(Photo: Marcus Lester)*

Marcus Lester, who took this photo, spent thirteen formative years working for Western Transport during which time he was nurtured and taught the job of truck driving by the late Arthur Gregory, seen here piloting BAE 986V down the M5 near Gordano Services in 1989. Arthur would apply the Boden trailer badge to the grille of all his Scanias at Western's to help him identify his unit in a fleet line up. On his retirement he gave the badge to his young charge, Marcus. BAE 986V, seen here pulling a smart Overlander trailer fitted with Freigat Bonnalack bodywork, ended its working days as a yard shunter and was finally sold for export to Africa along with its sister unit, BAE 985V. *(Photo: Marcus Lester)*

In these times of take-overs, amalgamations and large multinational companies it is always refreshing to see family-owned and run businesses doing well. Sparks Transport has grown steadily over the last seventy years or so and is now recognised as one of the UK's premier haulage fleets offering a top-quality service to both the domestic market and mainland Europe. Sparks has been using Scanias since 1968 and by the time the LB111 was introduced in 1975, the marque represented one hundred per cent of the twenty or so strong fleet. The situation remained the same for decades and only started changing in recent times. However, despite the odd Volvo and Renault, it is still Scanias that make up the backbone of this West Country fleet. PYC 271V was new in 1980 and remained with Sparks for a period of five to six years before disposal. Note the Seddon Atkinson 400 behind. Was this Britain's best response to the Swedish invasion? *(Photo: Marcus Lester)*

The exotically named Santa Fe Express ran a good number of LB Scanias - mainly LB81s in tractor and rigid form - on their storage and distribution work. However, they also ran a lone LB141 and a couple of LB111s. KAD 171V was the older of the two eleven-litre powered trucks and was marked out by this distinct variation on Santa Fe's eye-catching livery. In 1982 the company became part of the Oldacre Group and, as the new owners also owned a Mercedes-Benz dealership, the Cotswold fleet gradually changed over to MB products. *(Photo: Marcus Lester)*

Arthur Spriggs ran two LB Scanias. KBD 201V was the only eleven-litre example and started out in this fairly plain blue livery without the usual two-tone blue stripes. However, a contract for Smith's Containers saw the unit repainted in a red and white scheme along with around thirty other units in the fleet. During the repaint the number plate was moved up onto the grille and four spotlights were added under the bumper which was hazard striped black and white. Note the twin upright exhausts (later replaced by a single one) and the roof rack, which was accessed by a ladder mounted on the nearside of the cab. After many years of reliable service, including 38 tonne operation with impressive tri-axle super cube tilts, KBD 201V was sold and exported to Greece. *(Photo: Arthur Spriggs)*

VÄSTBO Transport AB was founded in Sweden in 1965, three years before the introduction of the original LB110, through the merger of several companies with transport histories that stretched back to the 1920s. This LB111 rigid was actually a wagon and drag combination, a type favoured by the company for its mixed traffic. The company's trucks were frequently featured in Scania's promotional material around this period. *(Photo: Scania AB)*

TSG 701V was Pollock's first LB Scania to be equipped with a full factory sleeper cab. In 1978 Scania made changes to the rear cab mounts and offered coil springs as an option. This went some way to redress the balance against new trucks like the Volvo F10/12 and Ford Transcontinental which featured advanced four-point cab suspension systems. Previously all LB cabs, once locked down, were virtually rigidly mounted and, without as much as a rubber block to absorb shock, the resulting ride was extremely firm. The coil-spring option also improved the longevity of the cab as it eliminated a stress point underneath which would ultimately fail once terminal rot had set in, breaking the structure's back. Another feature of the sleeper cab was extended lock/release handles (one each side) which would otherwise have remained far under the cab and thus be awkward to operate. Looking in fine condition when photographed in 1983, TSG 701V is seen changing trailers at the company's HQ in Olivebank. Note the absence of the traditional Pollock tartan on this example.

(Photo: John Henderson)

In contrast to Swains' French venture which utilised a localised version of their usual livery, this LB111, which operated out of Poole, featured an entirely different red and white livery. It was the only Swains vehicle based at Poole and the only one to appear painted in these colours. Note that this unit, in contrast to all Swains' other LB Scanias, was fitted with 88-gallon tanks and none of the badges were relocated for the livery.

(Photo: Swains of Stretton)

The late David Wilkie Haig started his haulage business in 1946 with an old Thornycroft lorry. Thirty years later he was running a mixed fleet of nearly forty units from various manufacturers including Atkinson, Leyland, Foden, Volvo and Seddon. International loads started in 1969 with an inaugural trip to Italy which became a regular weekly service. That first load was entrusted to an F88 240, which performed the task without problems. The international work grew steadily but it would be six years before the first Scania, an LB111, was bought. Two more followed, including LNS 38V seen here running at 38 tonnes in the mid 1980s. *(Photo: David Wakefield)*

This splendid LB111 was new to Ronnie Hudson in 1981. Ronnie, an owner-driver between 1973 and 1998, traded a T-registered LB141against RNG 242W with Morville Trucks of Lutterworth. Although a strong performer, the LB141 was dogged by a persistent whine from the back axle that was always most noticeable in the tunnels during Ronnie's frequent runs to Switzerland. When the problem was still not resolved after the unit's third trip to the dealer, Ronnie brokered a deal on the LB111 equipped with the 305 bhp DS11.02. Ronnie put the new unit to work supplying traction for both Norfolk Line and Matthews International until it was unfortunately stolen from Morville's yard in January 1983. As the LB111 range had been superseded by then, it was replaced with a new 112, EEX 617Y. The tasteful blue and white livery was not Ronnie's usual colours but that of the original buyer whose finance fell through before the sale. *(Photo: Adrian Cypher)*

This well-travelled LB111 of Richard Rooke Ltd is seen during a load transfer at the company's yard in North Yorkshire following an early morning return from London. SDS 723W was a frequent visitor to the capital's markets in the 1980s and as such carried a GLC exemption notice on the front grille. Note the rag attached to the mirror to prevent fouling and the condensation line on the big 88-gallon tank – a cold morning. *(Photo: Malcolm Slater)*

Charles Alexander ran a huge number of trucks (check the fleet number!) ranging from four- and six-wheel rigids by Ford and AEC through to premium tractor units including Leyland Marathons, Volvo F88s and many Scania LB111s. Many vehicles operated from here, the company's large depot in Old Ford Street, Aberdeen. General Haulage was handled alongside fish movement and a good selection of trailers was available from fridges to flats and curtainsiders. The simple but effective livery was instantly recognisable on the road and remained the same until 1986 when MANs also started to join the fleet. Alexander's finally closed its doors in the early 1990s. *(Photo: Andrew Keith)*

Testament to the adaptability of the LB111 is illustrated here with this LBS (6 x 2) solo rigid which earned its keep on medium-distance work taking pigs to market in the Netherlands. A D-Series Ford could have performed the same task but not with the same performance or comfort. The comparably large capital outlay for the Scania would have balanced out over many years of reliable service and a strong residual value would have been maintained. *(Photo: Scania AB)*

In Germany the LB111 was competing with some of the best-made trucks in Europe including those of the mighty Mercedes-Benz, the company that forced Scania to drop its original sprocket logo all those years before because of its similarity to the hallowed three-pointed star. Despite German loyalty to their domestic manufacturers, the LB111 sold well in this market even to own-account users as seen working here in this salt production facility. Note the Trilex wheels fitted to both tractors and trailers, the hub reduction axles and the unusual, angled mounting of the spare wheel. *(Photo: Scania AB)*

T & K Transport clearly had a connection with Yamaha motorcycles in the UK as their large fleet of LB111s all bore the Japanese manufacturer's name clearly in their headboards and, in at least one case, Team Yamaha logos on the cab too. Leigh Delamere Services was a regular rallying point for TK drivers where groups like this, with identical spec and all pulling semi low-loader trailers, would make an impressive sight. Vehicle movements using these trailers were a speciality of the firm and most of them could be extended on rams to accommodate different lengths and combinations of load. The aptly named 'Mellow Yellow' was barely a year old when photographed in 1982. *(Photo: Marcus Lester)*

This 1981 Essex-registered LB111 was purchased second-hand from an operator in Suffolk to swell the ranks of the Lydbury haulier, J Pugh and Sons. At the time the ten-strong fleet comprised entirely of sleeper cab LB111s that were employed solely on UK general haulage duties. The units worked into the 1990s and the company remained loyal to Scania until closing its doors in 2005. This impressive load of sawn timber, seen heading down the A30, was not a regular run but would have originated from Ramsford Mill. Note the company's fine, traditional-style livery. *(Photo: Marcus Lester)*

An interesting load carried in steel cages suggests a UK trip for OMY 942W on this occasion, but there is no disguising this unit's previous foreign adventures. Proudly wearing country stickers like passport stamps, this TIR LB111 shows evidence of visiting at least six countries on its travels, and the driver's penchant for one particular European capital is plain to see. Note the small locker, minus its door, between the battery box and the fuel tank. *(Photo: Marcus Lester)*

An atmospheric shot as a beautifully turned out LB111 of Andrew Wishart and Sons returns to the UK in 1982. The Wishart business was established before the First World War and, while maintaining a small size, it has become one of Scotland's most recognised fleets. Probably less than a year old at the time, JSP 466X would have been among the last of the LBs to be registered as the range was superseded in 1981 by the Series Two vehicles. However, this example shows that a 305 bhp powered LB111 was more than capable of hauling 38-tonne loads not just the length and breadth of the UK but also throughout Europe and beyond as well. Note how clean this example is on its return to the UK

(Photo: Adrian Cypher)

Currie European must be one of Scotland's biggest success stories in international road haulage. From humble beginnings in the early 1970s the company has grown into a massive concern which now runs over three hundred tractor units and six hundred trailers, still with essentially the same livery as this. Key to this success were the two European depots, one in Nijmegen and one in Gennevilliers, which the company set up. These helped to get return loads for the UK which eliminated costly empty running. Scania also played its part, as it was the reliability of trucks like this that helped to gain Currie its strong reputation.

(Photo: David Wakefield)

A friendly wave to the camera from the driver of this handsome LB111 as it drops into Dover. Geoff Gilbert built up his successful international operation with the help of Volvo F88s and F89s. However, a shift to Scania started in 1979 with the purchase of an LB111, WJL 902T. The unit impressed with more following until the balance tipped towards the Malmö product. The fleet is still mainly Scania today. KJL 629X was a late 305 bhp example and its gleaming paintwork shows how well the superb Gilbert livery looked when applied to the big cab. *(Photo: David Wakefield)*

This was one of two such outfits operated by Geoff Gilbert specifically to handle the movement of cut flowers from Holland and Italy. The loads were high cube and the wagon and drag combination offered a ten cubic feet advantage over the artic trailer alternative at the time. The LBS111s were powered by the DS11.02 version of the eleven-litre engine and featured a 6 x 2 chassis which, whilst over-spec for cut flower work, allowed the combinations to handle other loads too. For these vehicles, a variation on the familiar Geoff Gilbert livery was used incorporating a stripe for the dedicated bodywork and trailer. Note the spare wheel location on the near side. *(Photo: David Wakefield)*

This late example of an LB111 run by Bradshaws makes a pleasing combination as it travels along the M5 with one of the company's liveried curtainside trailers. Although this example is fully glazed, it was not uncommon, particularly on rigid chassis where rearward vision would often be obscured by bodywork, for the rear three-quarter windows to be deleted and sprayed over. If this was done at the point of ordering, then the apertures would remain sheeted over from the factory, but if done retrospectively it was common simply to paint over the glass. The latter method was easy to spot, as the window rubbers would remain, leaving a pronounced edge to the aperture instead of the depression of the factory option. Interestingly, when the rear window was deleted the factory would install a false metal panel, complete with rubber surround, in place of the glass. *(Photo: Marcus Lester)*

Looking superb in the once familiar chocolate brown and red livery of David Munro, this LB111 registered in 1981 was among the last of the breed to hit UK roads and as such was as good as an LB111 got. Examples of this vintage were gainfully working well into the 1990s before being exported for new careers abroad. Although all types of general haulage were undertaken, much of Munro's work involved the movement of paper from Fife. The company also ran a number of trucks in the silver and blue livery of main customer, Tullis Russell.
(Photo: John Henderson)

Scania LB141
King of the Road

By the LB140's seventh year in production the power race among super premium trucks was in full swing and the model was up against some serious competition for the top spot. Contenders from Mercedes-Benz and Fiat were creating a stir with big-capacity, low-stressed and naturally aspirated engines in the 350 bhp range while a host of trucks powered by Cummins' lusty, big-cam 14-litre NTC units (such as Ford's Transcontinental) were also making their presence known with similar outputs. The F89 (from home-grown competitors, Volvo) was introduced in 1970 one year after the LB140 and, as well as being a distinct thorn in the side, was something of an embarrassment for Scania as, although 20 bhp down on the LB140, the superb sixteen-speed SR61 gearbox gave the Gothenburg product the edge in both performance and fuel economy.

However, Scania had achieved its aims with the LB140 and, having introduced the industry to the big-power, low-stress concept, was now, while not about to re-write the book, certainly going to add a new chapter.

While retaining the same block and dimensions to achieve a higher power output, Scania made changes to a number of the engine's key components, notably the cylinder head which was redesigned along with the inlet ports. The latter acquired a rifled finish which caused a vortex in the combustion chamber to mix the fuel and air thoroughly. New pistons and valves were installed as well as camshafts featuring a revised profile and a higher capacity fuel pump to feed the injectors.

In the first major re-working of the DS14 engine, Scania's engineers managed to extract a further 25 bhp from the unit while increasing torque by an impressive 21 per cent to a new figure of 1102 lb/ft, now developed at a new lower speed of just 1300 rpm. The engine's overall speed was also reduced to 2000 rpm and a new economy zone was mapped at 1500 rpm (compared to the old engine at 1600 rpm). With even lower operating speeds than before and the same mild turbocharging as its predecessor, the new engine (designated DS14.01) was to prove hugely reliable and the 375 bhp it produced was enough to regain it the crown of 'most powerful truck in Europe' with a comfortable margin to spare.

Gearboxes, final drives and all other equipment remained as before and were the same as those offered on the LB111 and, as the LB140 had already been updated with all the same cab modifications from which the LB110 benefited back in 1973, there was little to do to the new model besides install the 141 model badge at the bottom of the grille.

Despite some initial indifference to the LB141, mostly because it took the 'let it lug' philosophy to a new level which many found hard to adjust to at first, Scania was back at the top and kept the model in production for five years during which time it produced a staggering 16,405 examples in the usual LB, LBS and LBT variants. Any initial doubts were soon reversed and the model went on to build a legendary reputation for performance and reliability with indestructible engines that were capable of staggering mileages even under the harshest of conditions.

This 1977 LBS141 made an impressive 24-metre outfit when operating in the Swedish timber forests on the bulk haulage of wood shavings. Conditions and turn-around times, rather than running weights, probably dictated this operator's decision when ordering the LBS141equipped with the new DS14 engine (now designated the DS14.01). It churned out a huge 375 bhp and, more significantly, 1102 lb/ft of torque at just 1300 rpm. This sort of performance put Scania back at the top of the power league and made the 141 an ideal choice for operations of this kind. *(Photo: Scania AB)*

The LBS chassis featured a double frame of great strength and flexibility with a uniform top surface that made it ideal for the mounting of bodywork, as is illustrated by this fine Swedish-registered example. This allowed a good potential for high-cube loads to be carried within normal truck dimensions and, when combined with the HB11 day cab option which had a short BBC (bumper to back of cab) measurement of just over 1.6 metres, it created a vehicle with enormous potential for moving bulk loads. *(Photo: Scania AB)*

Fred Hodgkins spent twenty-four years on the Middle East run, much of which was driving Grangewood's Scania LB140s and LB141s. Grangewood had started out serving the London markets with a small fleet of rigids but entered the big league and started operating to the Middle East through a connection with Christian Salvesen. Seventeen trucks were run from the company base in Greenwich (Grangewood was the telephone exchange for the area before STD codes were introduced) and, apart from four Volvos, all were Scanias. The trucks were painted in the colours of the Parachute Regiment, maroon and light blue, as the company founder and his drivers were all former members of that elite fighting force. Distant Baghdad was a frequent destination and Fred would spend thirteen weeks away at a time with the big Scania cab being his office and home. Loads were all temperature controlled and could be anything from chocolate to paint. Fred is pictured with his immaculate four-year-old LB141 on Mont Blanc in the early 1980s. *(Photo: Fred Hodgkins)*

Just like the LB140 before, Irish operators were quick to adopt the new LB141 following its launch in 1977 and found it perfectly suited to service the fast, refrigerated loads which they plied between the Emerald Isle, Great Britain and Europe. This beautifully presented example in the colours of the East-West fleet makes a stunning sight coupled with its matching trailer whilst on a break on the M4. Note the trailer axle spacing and robust spare wheel carrier. *(Photo: Adrian Cypher)*

VVW 910S started out in 1977 as an LB140, among the last registered in the UK with the updated front grille etc., but was re-badged in later life as an LB141. The unit was originally a 6 x 4 tractor employed on heavy haulage operations for Joseph Faulkes of Wednesfield. A substantial chassis stretch was required for the conversion which extended the wheelbase to rigid proportions before the high-capacity hydraulic lifting boom and bodywork could be mounted. The result was the extremely smart and very able recovery vehicle seen here at the Sandbach Services on the M6. Note the bumper modifications which acted to ballast the front end of the truck.

(Photo: Marcus Lester)

Sweden's manufacturing industries had developed at a pace through the twentieth century and enjoyed a worldwide reputation for quality. Scania was no exception; all major components were manufactured in house and one in ten of the workforce was employed just on inspection duties. All of which meant operators could sleep easily, safe in the knowledge that, even when operating at the extreme ends of haulage, their trucks were assembled with the greatest of care from the finest components available. *(Photo: Scania AB)*

Although the weight of this load, even with its lifting frame, would not have been great, the drag would no doubt have been a significant factor, especially at motorway speeds. Trueman Transport Services of Southampton did well to put their 1979 LB141, bearing the legend 'Ivor's truck', at the front of the twin-axle, trombone, flatbed trailer to move it. *(Photo: Adrian Cypher)*

John Brain's business started in the mid 1960s as an evolution of his father's company, Battens Transport. The rapid expansion enjoyed by Brain was in no small part due to the winning of a contract for ACT (Associated Container Transport) in the early days of the container revolution which alone employed over one hundred of Brain's white and yellow liveried trucks. At its peak the fleet swelled to over three hundred, operated mainly from the old Magnet depot in London Road, Thurrock. Despite the introduction of some Renault units in later years (Brain ran a dealership for the French manufacturer) Scanias formed the backbone of this massive Essex-based fleet right up to closure in the 1990s. Brain operated all sorts of Scania tractors from the work-a-day LB80 through to the mighty LB141. This fine LBS example joined the fleet in 1979 and is seen descending into Dover in the early 1980s. *(Photo: David Wakefield)*

The enormously successful Frigoscandia organisation grew from Sweden's early grasp of the principles of freezing common foods during the 1940s. Frigoscandia itself evolved from Helsingborgs Fryshus in the 1960s and, much like Scania, blossomed in that decade by providing an excellent product and service to an expanding market. Frigoscandia, as part of its integrated transport system, established state-of-the-art cold stores throughout Europe and also designed and built sophisticated freezing plants for customers around the world. Despite its advancing years this LB141, running at 38 tonnes with a smart Crane Fruehauf fridge, looks in superb condition as it exits the docks at Dover in the mid 1980s. *(Photo: David Wakefield)*

Battling a strong side wind as it heads down the M4, this 1979 LBS141 with 6 x 2 bogie-lift rear end was a left-hand drive Middle East veteran fitted with a host of extras to make life easier while working those demanding routes. These included 88-gallon tanks, roof rack, air-conditioning and, quite possibly, the de-luxe cab option. Interestingly, the fuel tank has been moved forward on this example and the battery box and air tank relocated elsewhere. Note the wide axle spacing of the unit's dedicated tilt trailer. *(Photo: Marcus Lester)*

Although not loaded, the heavy haulage potential of this 1979 LBT141 and low loader combination of Fenton-based Potteries Demolition is obvious. No doubt some of the demolition sites that YMF 700T had to negotiate made full use of the LBT's 6 x 4 layout with the three, air-controlled, diff locks coming in especially handy. The 'Gleason'-design bevel gear teeth within the axles were phosphated and case-hardened, making them exceptionally strong and reliable, ideal for this type of operation. Despite the nature of its work, YMF 700T wasn't, for operational reasons, equipped with the optional hub reduction. *(Photo: Marcus Lester)*

From the transportation of hay and specialist agricultural equipment to heavy haulage, since the late 1960s Hugh Wilson's business has gone from strength to strength as it has evolved and taken on new challenges. From humble beginnings with a second-hand rigid Bedford, the Suffolk fleet has grown into a state of the art international fleet. Wilson's earliest forays into heavy haulage were made possible with the acquisition in 1979 of this fine LB141, which was plated for 64-tonne operation. The big Scania joined LB110s and an LB111, which earned their keep mostly transporting combine harvesters around the UK and Europe. Being the star of the fleet, EGV 565T was never sold and in 2006 emerged from a long restoration. *(Photo: G. Wilson)*

Don't look down. The DS14.01 was in its element in mountainous regions such as this where its power and delivery came into their own. Scania's engineers had worked long and hard developing the LB141's distinct low-speed characteristics and geared it accordingly so that optimum engine speed was delivered around the 1500 rpm mark. As long as the driver was disciplined and, ignoring instinct, let the big engine lug in the higher gears, then remarkable fuel economy was possible while achieving fast journey times. Note the long rear overhang and trilex wheels of this beautifully turned-out LBS example. Maybe, while negotiating Alpine roads such as this, the optional safety window and mirror in the passenger door may have been desirable. *(Photo: Scania AB)*

Even if the going was flat the LB141 still made perfect sense for many operators, especially those running drawbar combinations where the drag from extra wheels, axles and bodywork would quickly take their toll on the fuel economy and journey times of trucks with lower outputs. Note the early-style air deflector and experimental sun visor fitted to this tidy Danish example photographed in 1979. *(Photo: Scania AB)*

This impressive LBT141 of Trendale Heavy Haulage was normally to be found moving all sorts of plant and machinery around the UK on low and semi low loader trailers. However, on this occasion it poses with this twin-axle flat, normally used for tractor movements, at the company's yard in Roche, Cornwall. The 150-tonne unit was bought second-hand from an operator in South Wales and proved to be very capable of the STGO CAT 2 operations to which it was assigned. Note the gap in the chassis as neither a second diesel tank nor a spare wheel carrier was fitted. *(Photo: Marcus Lester)*

Highway Commercials of St Blazey had run some interesting wreckers since the business was started in 1963 including a converted, ex-War Department, 6 x 6 Scammell tank transporter with petrol engine and a Leyland Super Mastiff with Perkins V8 diesel. So, when FGB 350T with its big V8 and hefty 150 tonne rating came up for sale locally in 1989, they snapped it up for conversion into a wrecker. Highway Commercials, who specialised in building and repairing truck bodies, fitted the unit with substantial TFL lifting gear taken from an old F88, and built the surrounding bodywork from scratch. Undaunted by the severe hills surrounding the company's base, FGB 350T performed recoveries, including fully laden artics, without the slightest protest for over six years before eventually being sold on.
(Photo: Marcus Lester)

Trevor Caithers was another Irish operator who favoured the LB141 for express, TIR, fridge work. Here, one example with the familiar TC livery, is seen parked up in 1982. Because Scania was able to keep pace with other manufacturers and keep the DS14 at the forefront in engine output it was common for those who had operated the big V8 in the LB140 and LB141 to progress through to the R142 and subsequently the R143. The durability of the LB141 was such that it was not unusual to find examples still working in the same fleet alongside both the later types, especially as some drivers were loath to give up their beloved 141s. *(Photo: Marcus Lester)*

Bro wrecker equipment exploded out of Sweden around the same time as the original LB140 in 1969 and was met with similar approval by an industry in need of innovative solutions. The small company from Kristinehamn near Gothenburg had roots that went way back to 1945 but, by marketing its top-quality products in conjunction with Scania, it was able to reach a world market. Quality was second to none, as was the ergonomics of the design that enclosed everything within neat and functional bodywork. Hydraulic lifting booms and winches (the latter remote controlled) had the capacity to recover a 54-tonne plus combination with ease. *(Photo: Scania AB)*

Even as a 4 x 2 tractor, the LB141 made a steady platform for difficult loads such as liquids. The various spring options including air, parabolic and multi-leaf could be enhanced with stabilisers (anti-roll bars) and load-sensing brake valves. This Swedish-registered example demonstrates the unit's superb traction in tricky conditions, even when part-loaded, as it negotiates a frost-covered rise. It was also in conditions like this that the heated driver's seat came into its own. *(Photo: Scania AB)*

Kevin Latham, now Middle East Development Manager for Astran, must have been one of the few lucky enough to learn to drive with an LB141. KFJ 730V is believed to have started out with CATS (Chard and Axminster Transport) and at some point acquired a thirteen-speed fuller gearbox which improved shift time and gave a slightly higher top speed. On passing his test, Kevin was given Gordon Smith's other LB141 as his regular drive. This was a later unit registered on a W plate and had originally belonged to Mortimers of Wiltshire who used it for potato haulage. Gordon Smith continued to run a 141 until his death in 1996. Note that the Scania badge has been removed and replaced with old-style lettering. *(Photo: Kevin Latham)*

Thrapston Warehousing didn't run any trucks of its own besides yard shunters, and instead used sub-contractors, many of which were owner-operators. This smart 1979 LB141 running in Thrapston's own colours was probably that of owner-driver Dick Craddock. The tri-axle, step frame tilt was TW's favoured trailer type and was ideally suited to the delivery of the split, groupage type of loads which the company specialised in tipping throughout Europe and the UK. *(Photo: David Wakefield)*

Because of its flagship status, the LB141 would often enjoy extra attention from its owners and drivers. This fine example features a beautifully detailed livery and many useful accessories such as the roof rack and chassis locker. In an attempt to improve economy, a neat under-bumper air-dam has been fitted to smooth the flow of air under the tractor. LAT 904V's driver, obviously proud of his DS14 power plant, has also fitted additional V8 badges and an old 'Super' logo to emphasise the point. Note the clever homage to the Swedish flag incorporated in the bumper. *(Photo: Dave Wakefield)*

The huge, low-down torque of the DS14 was perfect for getting heavy loads moving, and the rugged chassis and drive train could absorb enormously punishing loadings. Econofreight were one of the 'big name' companies that found the big Scania's attributes well suited to their heavy haulage operations. This 6 x 4 version was an ex-Sunters unit and is seen moving a large railway carriage under-frame on an extending semi low-loader trailer in the mid 1980s. *(Photo: Adrian Cypher)*

This 1981 LB141 demonstrates again that the big Scania could earn its keep just as effectively as a lower powered unit while working at the UK's old 32-ton limit. At the time 250 bhp would have been deemed perfectly adequate by most operators and 290 bhp a positive luxury. Considering that the world was recovering from a fuel crisis when the LB141 was introduced, and Great Britain recovering from a bleak economic period on top of that, the willingness of UK operators to employ the 375 bhp machine at 32 tons proved that Scania's low-speed, high-torque policy really did work and that good fuel economy was possible from a large-capacity engine. *(Photo: Scania AB)*

ROBA's charismatic MD, Seymour Grann, was a Scania man through and through and swore by the marque for the company's scheduled runs to Switzerland. Four LB141s were operated from the company's base in Harlow alongside LB111s and a few LB81s. Other makes were tried from time to time but all disappointed, including Volvo. Essex-registered ROO 110W was a typical example of the ROBA fleet at the start of the 1980s with a 4 x 2 chassis and left-hand drive. It is seen here dropping into Dover at the start of another trip to Basle. Later, R142s would join the fleet, continuing the Scania V8 bloodline for ROBA. *(Photo: Dave Wakefield)*

This dramatic backdrop of snow-dusted mountains which often represented the final destination for ROO 110W, some twelve hundred miles from home and parked outside ROBA's Basle HQ. Further depots were maintained in Italy, Spain, Portugal and Dover. However, the trucks would visit all European countries with the groupage loads, which could vary from a few tonnes to a full load. The Scanias were very popular with the ROBA drivers because of their strong reliability and the knowledge that if there was a problem help was never far away thanks to Scania's extensive service network throughout Europe. *(Photo: ROBA)*

Paul Hart ran this splendid LBS141 for thirteen years as an owner-driver during which time the truck performed faultlessly. Although it was originally registered in the south, Paul actually bought the unit from a company in Manchester as a 4 x 2 LB141. It was later converted with a tag lift which made it more versatile when pulling other people's trailers. The unit is pictured performing a one-off job for Paul's father who ran a marquee business. The purpose-built York trailer was 36 foot, the biggest size that could negotiate the yard. *(Photo: Paul Hart)*

Pegler & Louden bought this LB141 new in 1981 to service their own business, transporting industrial valves to depots around the UK. Jim Low was the lucky driver given charge of the vehicle at the tender age of just twenty-two. Jim would leave Pegler's Glasgow base on a Monday fully loaded and stay out until Friday. The unit was serviced every four weeks. When it was replaced at three years old it had covered 350,000 km, still had its original clutch and the only replacement part, other than service items, was a new alternator. *(Photo: Jim Low)*

Some minor damage to the front offside spoils this otherwise smart LB141 of MacHayes International Transport, PPW 156W. Being a 1980-81 unit, it was built towards the end of the LB production and seems well-equipped with air conditioning, roof rack, extra low-level door mirror, sun visor and additional driving lights - in fact everything you would expect of a TIR spec machine. Although all these features were available individually as extras, in all probability this cab was actually the de-luxe version which incorporated them as standard. Note the minimal clearance of the trailer on this example as it moves what looks like an ex-MOD bulldozer in the mid 1980s. *(Photo: Adrian Cypher)*

When Michael L Faiers set himself up as an owner-driver back in 1978 he fancied a Scania LB141 but, with demand for the model so high, he found a six-month waiting list. A DAF 2800 nobly filled the gap, proving a good truck, and even afforded Michael priority unloading on his frequent loads to the manufacturer's Eindhoven factory. However, it wasn't too long before the first LB141, YGV 80S, was bought and not long after that NRT 890W doubled the size of Michael's fleet. With a driver employed for the older unit, Michael took the wheel of his new charge, piloting it on two continental trips a week to Belgium and Holland, mainly for Nipress. The truck was loaded with extras such as two sets of spotlights, quadruple fog lamps, double 88-gallon tanks, a continental style headboard, sun visor, wheel trims and roof spoiler, all of which pushed the price up to an eye-watering £27,782. However, the unit made a stunning sight, was a great advertisement for Michael and realised a strong residual when sold two years later. Michael stuck mainly with V8 Scanias until he retired in 2006 at which point he had owned and operated up to forty examples. NRT 890W is pictured waiting to unload at a cash and carry in Wales when just six months old. Note the electrical cable running down the trailer side which powered a removable lightboard hung on the rear as the Nipress tri-axle tilt had no rear lights of its own. *(Photo: Marcus Lester)*

Cadnam-based Waters Bros has over thirty-five years' experience of hauling timber out of the New Forest and from early on it has found Scanias the ideal tool for the job, offering the perfect mix of on- and off-road abilities. FHO 921W was originally a 4 x 2 unit but Waters had it converted with a Granning air-lift fitted by Andover Trailers. Waters, with a second-hand buying policy, would scout local dealers for suitable used LB Scanias and, with nights out extremely rare, would generally pick day cabs, making FHO 921W unusual in the fleet at that time. The majority of loads for FHO 921W were destined for the timber mill of RS Giddings where they would become fence panels and posts. *(Photo: Phil Waters)*

Ernie Felgate had worked for both Russell Davies and Brain Haulage, so it was no great surprise that he opted for Scanias when he set up on his own with a batch of LB141s leased through ScanTruck. A good contact in Italy resulted in the smartly liveried trucks providing traction for Merzario road/rail trailers. As Ernie had married John Brain's sister, this work also filtered through to the Brain fleet. *(Photo: David Wakefield)*

This tidy 1980 LB141of HiTrans International Freight with its tri-axle tilt is representative of express TIR outfits throughout Europe in the late 1970s and 1980s. Apart from the right-hand drive and UK registration, it could easily have been from Holland, Germany, France, Italy or any country in the EEC - such was the popularity of Scania's flagship. Note the chassis locker fitted to this example with a 3.4 metre wheelbase. *(Photo: David Wakefield)*

Robin East, former Transport Manager for Bejam, established Rokold International (later European) in 1976 with a brand new DAF 2800 DKS. From the start his policy was to buy new and replace often, usually every two years. The 2800 DKS proved reliable with good performance and DAF trucks were to form the backbone of the immaculate blue and white fleet, which peaked at around fifty units, for many years after. Indeed, other manufacturers only got a look in when the decision to operate three-plus-three axle combinations ruled out DAF as the trucks carried a big weight penalty over rivals at the time. ROO 112W was therefore something of a rarity and actually came to Rokold as part of the deal when Rokold bought out the business of former sub-contractor, John Cooper, in the early 1980s. During its two-year stint, the big Scania performed deliveries throughout Europe without any problems and actually undertook Rokold's first load to Romania. A single LB111, also part of the deal, was operated on a dedicated shuttle run to and from Belgium with loads of frozen chips. Robin sold his company in 1998 to Wincanton by which time the fleet mostly comprised Scanias with a few Volvo and Mercedes to make up the numbers. *(Photo: David Wakefield)*

One of the last of the line, XAR 927X, in a fine state of preservation, seems to have found a niche in later life as the prime mover for this vintage steam-roller. Using an old fashioned, four-in-line low loader trailer, shod with military-pattern tyres, the unit moved the veteran machine around the country for steam fairs and rallies. Note the spray suppression equipment fitted in the wheel-arch of this example. *(Photo: Adrian Cypher)*

Another sad loss to the haulage industry was Andy's of Banbury. This company was an early user of LB Scanias and its smartly painted trucks would often appear in Scania brochures and literature during the 1970s. This late example of an LB141 was clearly well looked after by its driver and features typical period accessories in the form of extra running lights, sun visor and wheel trims. Note Andy's clever, all-direction, big 'A' logo. *(Photo: Marcus Lester)*

Although rarely seen in Europe, this interesting comparison shows the marked difference between the standard LB chassis and that of the LK with a front axle mounted 415 mm further forward. The range was available with the same chassis/drive layouts as the regular LB with model designations of LK (4 x 2), LKS (6 x 2) and LKT (6 x 4). Because the air filter housing location at the front of the chassis was lost, LKs featured a snorkel type air stack mounted up the back of the cab; the first time Scania used such a device on a forward control truck. *(Photo: Scania AB)*